Macmillan Computer Science Series
Consulting Editor
Professor F. H. Sumner, University of Manchester

S. T. Allworth and R. N. Zobel, *Introduction to Real-time Software Design, second edition*
Ian O. Angell and Gareth Griffith, *High-resolution Computer Graphics Using FORTRAN 77*
Ian O. Angell and Gareth Griffith, *High-resolution Computer Graphics Using Pascal*
M. Azmoodeh, *Abstract Data Types and Algorithms*
C. Bamford and P. Curran, *Data Structures, Files and Databases*
Philip Barker, *Author Languages for CAL*
A. N. Barrett and A. L. Mackay, *Spatial Structure and the Microcomputer*
R. E. Berry and B. A. E. Meekings, *A Book on C*
G. M. Birtwistle, *Discrete Event Modelling on Simula*
T. B. Boffey, *Graph Theory in Operations Research*
Richard Bornat, *Understanding and Writing Compilers*
Linda E. M. Brackenbury, *Design of VLSI Systems – A Practical Introduction*
J. K. Buckle, *Software Configuration Management*
W. D. Burnham and A. R. Hall, *Prolog Programming and Applications*
J. C. Cluley, *Interfacing to Microprocessors*
Robert Cole, *Computer Communications, second edition*
Derek Coleman, *A Structured Programming Approach to Data*
Andrew J. T. Colin, *Fundamentals of Computer Science*
Andrew J. T. Colin, *Programming and Problem-solving in Algol 68*
S. M. Deen, *Fundamentals of Data Base Systems*
S. M. Deen, *Principles and Practice of Database Systems*
Tim Denvir, *Introduction to Discrete Mathematics for Software Engineering*
P. M. Dew and K. R. James, *Introduction to Numerical Computation in Pascal*
M. R. M. Dunsmuir and G. J. Davies, *Programming the UNIX System*
K. C. E. Gee, *Introduction to Local Area Computer Networks*
J. B. Gosling, *Design of Arithmetic Units for Digital Computers*
Roger Hutty, *Fortran for Students*
Roger Hutty, *Z80 Assembly Language Programming for Students*
Roland N. Ibbett, *The Architecture of High Performance Computers*
Patrick Jaulent, *The 68000 – Hardware and Software*
J. M. King and J. P. Pardoe, *Program Design Using JSP – A Practical Introduction*
H. Kopetz, *Software Reliability*
E. V. Krishnamurthy, *Introductory Theory of Computer Science*
V. P. Lane, *Security of Computer Based Information Systems*
Graham Lee, *From Hardware to Software – an introduction to computers*
A. M. Lister, *Fundamentals of Operating Systems, third edition*
G. P. McKeown and V. J. Rayward-Smith, *Mathematics for Computing*
Brian Meek, *Fortran, PL/1 and the Algols*
Barry Morrell and Peter Whittle, *CP/M 80 Programmer's Guide*
Derrick Morris, *System Programming Based on the PDP11*
Pim Oets, *MS-DOS and PC-DOS – A Practical Guide*
Christian Queinnec, *LISP*
E. R. Redfern, *Introduction to Pascal for Computational Mathematics*
Gordon Reece, *Microcomputer Modelling by Finite Differences*
W. P. Salman, O. Tisserand and B. Toulout, *FORTH*
L. E. Scales, *Introduction to Non-linear Optimization*
Peter S. Sell, *Expert Systems – A Practical Introduction*
Colin J. Theaker and Graham R. Brookes, *A Practical Course on Operating Systems*
J-M. Trio, *8086–8088 Architecture and Programming*
M. J. Usher, *Information Theory for Information Technologists*
B. S. Walker, *Understanding Microprocessors*
Peter J. L. Wallis, *Portable Programming*
Colin Walls, *Programming Dedicated Microprocessors*

0 828 239 0

I. R. Wilson and A. M. Addyman, *A Practical Introduction to Pascal – with BS6192, second edition*

Non-series
Roy Anderson, *Management, Information Systems and Computers*
J. E. Bingham and G. W. P. Davies, *A Handbook of Systems Analysis, second edition*
J. E. Bingham and G. W. P. Davies, *Planning for Data Communications*

Abstract Data Types and Algorithms

Manoochehr Azmoodeh

Computer Science Department,
University of Essex,
Wivenhoe Park, Colchester

MACMILLAN
EDUCATION

First published 1988

Published by
MACMILLAN EDUCATION LTD
Houndmills, Basingstoke, Hampshire RG21 2XS
and London
Companies and representatives
throughout the world

Typeset by
TecSet Ltd, Wallington Surrey

Printed in Hong Kong

British Library Cataloguing in Publication Data
Azmoodeh, Manoochehr
    Abstract data types and algorithms.—
    (Macmillan computer science series).
    1. Electronic digital computers—
    Programming   2. Algorithms
    I. Title
    005.1   QA76.6

    ISBN 0-333-42127-2
    ISBN 0-333-42128-0 Pbk

To My Parents

Contents

Preface

This book is intended as a second course on programming with data structures. The approach taken is based on the notion of an Abstract Data Type which is defined as an abstract mathematical model with a defined set of operations. The treatment of abstract data types is very informal. The specification of data types and their corresponding operations are presented in a form directly representable in a Pascal-like language.

The primary advantage gained by using abstract data types in program design is that they lead to better structured and modular programs. This type of modularity ensures that logical tasks and data are grouped together to form an independent module whose functional specification is visible to other programs using it. These application programs do not require detailed information on how the tasks and data are realised in computer hardware. Thus various strategies may be used to implement a module's data types and operations so far as they conform to the specification of that module. This separation of specification and implementation of programs ensure that many design decisions, such as efficiency, trade-offs between speed and storage etc., can be made at a later stage when the logical properties of data types and their operations have proved to satisfy the application.

In Part I, abstract data types are classified with respect to the single relation 'predecessor/successor' which may or may not exist among the individual elements of a data type. Those data types for which this relation does not hold are primarily sets. Those with this relation defined on them are further classified into linear and non-linear data types. Non-linear data types are finally classified into trees and graphs. Only when the specification of an abstract data type is completely given do we begin to consider implementation issues. In this phase, the major emphasis rests on 'efficient' implementations of a data type and its operations. Therefore, a chapter is devoted to discussing the notion of 'efficiency' of an algorithm so that later in the book the efficiency of the implementations of abstract data types and algorithms may be quantified. Furthermore, this enables us to compare different implementations of the same abstract data type or the same task in a given programming language and a computer system.

Prerequisites

The reader is assumed to have practical experience of a Pascal-like language. Although Pascal is used to express all the algorithms, they could easily be transformed into other similar languages. The efficiency of the algorithms is a major concern of the book and therefore it is advantageous if the reader has some knowledge about how various constructs of a Pascal-like language are implemented – with a view to comparing their relative speeds. However, this is not essential since the brief explanations given in the book should be sufficient to grasp these ideas.

Structure of the Book

The book is in two parts. Part I begins by examining the time and space requirements of computer algorithms and develops a notation which is used in the remainder of the book to compare various implementations of abstract data types. Chapter 2 introduces the concept of abstraction in program design and shows how data types can be abstracted from their detailed implementation issues. Section 2.3 briefly describes a method of formally specifying the data type stack and proving that its implementation conforms to that specification. The remainder of the book relies on informal specifications and proofs. This section and the bibliographic notes at the end of chapter 3 are intended as a guide for more interested readers, should they wish to explore the topic more rigorously; therefore it may be skipped in its entirety or on the first reading. Chapters 3 to 7 then discuss a variety of abstract data types including stacks, queues, lists, sets, trees, relations and graphs. Each abstract data type is defined as a mathematical model with a number of operations defined on that model. The operations are given as a set of Pascal procedures and functions. Various implementation strategies of these data types are discussed in terms of Pascal constructs. Examples of applications of these abstract data types are also given.

Part II further describes many algorithms and common techniques for developing efficient algorithms using abstract data types. Programming paradigms such as divide and conquer, dynamic programming, graph searching, tabulation techniques and randomised algorithms, are discussed. Chapter 9 develops a divide and conquer algorithm for the sorting problem and it is shown how implementation considerations are delayed until the final stage of program design. Finally, it is emphasised that these algorithm design techniques are not universal tools for producing efficient algorithms for all problems. In particular, chapter 12 briefly discusses the class of 'hard' problems for which most algorithms are not very efficient. A few techniques are described, such as how to handle such inherently difficult problems.

Exercises and Assignments

Computer programming can only really be learnt through experience. Merely reading about it is no substitute for first-hand practical experience. Therefore, to reinforce the ideas, exercises are included at the end of each chapter. These are usually simple and can be tackled after the content of the chapter is well understood. Some involve verifying algorithms by checking them manually on paper, while others require the writing of programs or fragments of programs. Guidance on the solutions of selected exercises is provided at the end of the book. Since the emphasis of the book is on the efficient implementation of various abstract data types and the comparison of algorithms, a few practical assignments are also included among the exercises. These take the form of designing, writing and testing actual programs. A few guidelines on the design of data types and algorithms for these assignments are given in the solutions to exercises.

Bibliographic Notes and Further Reading

No book of this size can deal comprehensively with abstract data types, data structures, algorithm design and the theory and practice of programming with data. However, the material presented here should provide an introduction to the practical issues of abstract data type and algorithm design. For motivated readers, each chapter concludes with a brief list of related literature and some comments on it. Many of the references are articles published in computer journals; these journals and their abbreviations are given below.

ACM: Association for Computing Machinery
ACM Computing Surveys
ACM TODS: ACM Transactions on Databases
Acta Informatica
BIT: Behandlings Informations Tidskrift – Denmark
CACM: Communication of the ACM
Computer Journal – British Computer Society
Journal of the ACM
SIAM: Social and Industrial Applied Mathematics
SIAM Journal of Computing
Software Practice and Experience

Acknowledgements

My sincerest thanks go to Martin Henson for many stimulating discussions. He also read the drafts of some chapters and I found his constructive comments and criticisms invaluable.

I acknowledge the assistance of the Department of Computer Science at the University of Essex in providing computer facilities and support while this book was being produced.

I thank all the students attending the course on data structures and algorithms at the University of Essex who provided me with many useful comments on the material presented in the book.

I also greatly appreciate all the help and effort that the secretaries, Marisa Bostock and Ann Cook, put into the typing of the first draft.

And last but not least, I am most grateful to my wife, Hengameh, who was my main source of encouragement whenever my enthusiasm waned, and for her continued support and forbearance.

Colchester   Manoochehr Azmoodeh
1987

Preliminaries and Notation

The only prerequisites for reading this book are practical experience in programming in a Pascal-like language and a sound understanding of general programming concepts together with a good grasp of Pascal-like data types (that is, reals, integers, characters, arrays and records). However, the analysis of many data structures and algorithms in the book requires a minimal understanding of such basic mathematical concepts as sets, functions etc. For the purposes of reference and also to introduce the notation used, a brief survey of these concepts is given below. A discussion of the idiosyncrasies of Pascal syntax, as used in the book, then follows.

Mathematical Backgrounds

Sets

A *set* is a collection of objects called *members*. The members of a set are usually of the same type. For example

S_1 : A set of colours = {red, blue, yellow, white}
S_2 : The set of negative numbers = {x is an integer and $x < 0$ }
S_3 : An empty set = { }.

The '{' and '}' are used to represent sets; in S_2, x is a variable and '|' should read as 'such that'. Therefore set S_2 should read as 'the set of all values of x such that x is an integer and x is less than 0'. 'A is a member of a set S' and 'B is not a member of a set S' are denoted as: $A \in S$ and $B \notin S$. Thus

red $\in S_1$ brown $\notin S_1$
$-15 \in S_2$ 3398 $\notin S_2$

Members of sets are not ordered. Therefore {red, blue} = {blue, red}. The number of members in a set S is called its *cardinality* and is denoted by $|S|$. Therefore, $|S_1| = 4$, $|S_3| = 0$.

Subsets

Set S is a subset of S$'$ if every member of S is also a member of S$'$. This is denoted as $S \subset S'$. We denote 'S is not a subset of S$''$ as $S \not\subset S'$.

{red, blue} $\subset S_1$
{−23, −345, −9999, −12} $\subset S_2$
{red, white, black} $\not\subset S_1$
{red, yellow} $\subset$ {red, yellow}

Intersection

Set S is the intersection of sets S' and S'' if every member of S is a member of S' and also a member of S''. This is denoted as: $S = S' \cap S''$.

{red, blue} = {red, yellow, blue, white} $\cap$ {blue, black, red}

Union

Set S is the union of sets S' and S'' if every member of S is either a member of S' or it is a member of S'' or it is a member of both S' and S''. This is denoted as: $S = S' \cup S''$.

{red, blue, white, black} = {red, blue, white} $\cup$ {blue, black, red}

Set Difference

Set S is the set difference of sets S' and S'' if every member of S is a member of S' but is not a member of S''. This is denoted as: $S = S' - S''$.

{red, blue} = {red, yellow, blue, white} − {black, yellow, white}

Cartesian Product

The cartesian product of two sets S and S' is defined as a set of *ordered pairs* (x,y) $((x,y) \neq (y,x))$ such that x is a member of S and y is a member of S'; that is

$$S * S' = \{ (x,y) \mid x \in S \text{ and } y \in S' \}.$$

For instance

{red, blue} * {4, 8} = {(red, 4), (red, 8), (blue, 4), (blue, 8)}

It can be shown that $|S * S'| = |S| \times |S'|$ for finite sets. Similarly, the cartesian product of n sets $S_1, S_2, \ldots, S_n$ can be defined as

$$S_1 * S_2 * \ldots * S_n = \{(a_1, a_2, \ldots, a_n) \mid a_i \in S_i \text{ for every } i\}$$

Each $(v_1, v_2, \ldots, v_n) \in S_1 * S_2 * \ldots * S_n$ is called an *n-ary tuple*.

Sets of Sets

Sets can be members of other sets. For instance, a set of families is a set of sets where each member set is the names of people in a family:

   { {Fred, Mary, John}, {Alf, Claire, Sue, Joe}, . . . , {Jim, Anne} }

Relations

A relation R on S_1, S_2, . . ., S_n is defined as a subset of the cartesian product of these sets:

   $T \subset S_1 * S_2 * \ldots * S_n$

R is called an *n-ary* relation. When $n = 2$, R is called a *binary relation*. When $n = 3$, R is called a *ternary relation*. For example

   relation age = { (Fred, 25), (Jim, 16), (Mary, 25) }

where S_1 = {Fred, Jim, Mary} and S_2 = {25, 16}.

Mappings (or Functions)

A *mapping* or a *function*, F, from set S_1 to set S_2 is a binary relation on S_1 and S_2 such that for every element x in S_1 there is at most one element y in S_2 such that $(x,y) \in F$. This is denoted as $F(x) = y$. In other words, if $F(x) = y$ and $F(x) = z$, then y must be equal to z. For instance, age is a function from the set {Fred, Jim, Mary} to the set {16, 25}. Set S_1 is called the *domain* of the mapping F and set S_2 is called the *range* of F. This is usually represented as

   $F : S_1 \to S_2$

Thus

   age : {Fred, Jim, Mary} $\to$ {16, 25}

   An *injective* or *one-to-one* mapping, F, is one such that $F(a) = F(b)$ if and only if $a = b$. If F is not injective, it is called a *many-to-one* function.
   A *surjective* mapping, F, is one such that for each $y \in S_2$, there exists an $x \in S_1$ such that $F(x) = y$.
   A *bijective* mapping, F, is one which is both injective and surjective.

Permutation and Combinations

A *permutation* of m-ary tuples on a set $S = \{a_1, a_2, \ldots, a_n\}$ is a set of *ordered* tuples $(x_1, x_2, \ldots, x_m)$ such that $x_i \in S$ (for all i) and $x_i \neq x_j$ if $i \neq j$.
   A *combination* of m-ary tuples on a set $S = \{a_1, a_2, \ldots, a_n\}$ is a set of *unordered* tuples $(x_1, x_2, \ldots, x_m)$ such that $x_i \in S$, $x_i \neq x_j$ if $i \neq j$. For example, for $S = \{1, 2, 3\}$ and $m = 2$:

Permutation on S = { (1, 2), (1, 3), (2, 3), (2, 1), (3, 1), (3, 2) }
Combination on S = { (1, 2), (1, 3), (2, 3)}

The number of permutations of m values out of n values is denoted by P_m^n and the number of combinations of m values out of n values is denoted by C_m^n. It can be shown that $P_m^n = n!/(n-m)!$ and $C_m^n = n!/(n-m)!m!$.

Sequences

A *sequence* is a linear and ordered list of objects usually of the same type: $a_1, a_2, a_3, \ldots, a_n$. A *sub-sequence* of this sequence is a sequence: $a_{i_1}, a_{i_2}, \ldots, a_{i_m}$ where $(i_1, i_2, \ldots, i_m)$ is a combination of m values from $\{1, 2, \ldots, n\}$ such that $i_j > i_k$ if $j > k$.

Logical Operators and Quantifiers

A *n-ary* *predicate* $P(a_1, a_2, \ldots, a_n)$ is a property which is true for some *n*-ary tuples and false for other tuples. For example

person(Mary) is true
person(chair) is false
age(Mary, 40) is true
age(Mary, 54) is false

Predicates can be combined to form complex properties by using logical operations: and, or, not, ⇒ (implies), ≡ (equivalent). Properties can have *quantifiers*: *for all* ∀ and *there exists* ∃. For example

$\forall x$ number$(x) \Rightarrow$ positive$(x \times x+1)$: For all x, if x is a number, its square plus 1 is a positive number

$\forall x \exists y$ greater(x, y): For all x, there exists a value of y such that $x > y$

Programming Language: Notations and Assumptions

The language used in this book is primarily standard Pascal with a few minor variations. However, the algorithms can be expressed in Modula-2 or any other similar language with little difficulty. Now a few examples of the notation and assumptions made in the book regarding Pascal programs will be given.

Throughout the book the existence of the procedure error is assumed:

procedure error;
begin

```
        writeln('There has been an error.');
        goto 999
end;
```

999 is assumed to be a label at the end of the program. This procedure is called every time an error condition has occurred and therefore the program is terminated. In a few places, this procedure is explicitly given a parameter to describe the nature of the error that has occurred.

The data type 'elemtype' is used in the book and refers to a Pascal definable type. The assignment operation is supported for variables of type 'elemtype'. Usually 'elemtype' is used as the type of the individual elements of a structure. In a few places elemtype can only refer to a record structure of Pascal, but this is evident from the context. When appropriate, the values of type elemtype are assumed to be ordered such that the relations '$>$', '$<$' etc. can be applied to them.

The identifiers used in the Pascal programs are allowed to have underscore characters '_' and also may have subscripts or superscripts. (This is intended for improving the readability of programs. To run these programs, such decorations should be omitted.)

```
var a_or_b : boolean;
    a₁, a₂ : integer;
```

A major notational extension is the use of an abstract control construct. This is primarily a means for traversing all the members of a set:

for all $s \in S$ **do** P(s);

where S is a set and P is a procedure. This statement is equivalent to executing P for every member of the set S. In Pascal, this control structure will be implemented as

procedure process_all(**var** S : zet; **procedure** P(**var** s : elemtype));
(* This procedure would execute P(s) for all elements s in the set S *)

. . . .

process_all(S, P);

Note that the type 'zet' refers to abstract data type 'set' as discussed in chapter 5. Pascal 'set' data types are not allowed to participate as variable parameters of sub-programs. A similar control construct will be used for traversing members of a relation R:

for all $(a,x) \in R$ **do** P(x);

where a is an element (constant) and x is a variable. This is equivalent to executing P for every x such that the pair (a,x) is in R.

Finally, the symbols '{' and '}' are used to denote sets and the symbols '(*' and '*)' are exclusively used to denote comments in Pascal programs.

PART I:
DESIGN AND ANALYSIS
OF DATA TYPES

1 The Complexity of Algorithms

Usually there are many programs or algorithms which can compute the solution of a specified task or problem. It is necessary, therefore, to consider those criteria which can be used to decide the best choice of program in various circumstances. These criteria might include such properties of programs as good documentation, evolvability, portability and so on. Some of these can be analysed quantitatively and rigorously, but in general many of them cannot be evaluated precisely. Perhaps the most important criteria (after 'correctness', which we will take for granted) is what is loosely known as 'program efficiency'. Roughly speaking, the efficiency of a program is a measure of the resources of time and space (memory or store) which are required for its execution. It is desirable, of course, to minimise these quantities as far as possible, but before we can consider this it is necessary to develop a rigorous notation which can be used as a yardstick for our programs. It is the purpose of this chapter to introduce such a notation, along with some illustrative examples.

1.1 Comparing the Efficiency of Algorithms

Let us consider a few examples that illustrate the differences in time and space requirements of different programs for a particular problem. Consider the following search problem:

> The array $A[1..128]$ contains 128 integers in increasing order of magnitude (with no duplicates). Write a program to find the index i such that $A[i] = C$ where C is a given integer constant. If there is no such i, a zero should be returned.

A simple solution may start scanning the array from left to right (in ascending order of index) until either a location is found which contains C or the end of the array is reached (see function search on page 3).

Intuitively, it can be seen that, in the worst case, this sequential search may have to inspect each element of A. Now we can use the fact that the array is ordered to develop a new algorithm. We compare C with the middle element of the array $A[64]$ (in this case). If $C = A[64]$, the solution is found. If $C < A[64]$ then, if present, C must be in $A[1]...A[63]$ and if $C > A[64]$, it must be in $A[65]...A[128]$. In either case, a search within a smaller sub-array should be initiated. Thus the function b_search can be developed:

```
const N = 128;
    . . . .
function search (C : integer) : integer;
  var J : integer;
  begin
    J := 1;    (* initialise J *)
    (* while end of list is not reached and C
       is not found, continue the search *)
    while  (A[J] < C) and (J < N) do
      J := J + 1;
    if A[J] = C then
        search := J    (* ... C is found and A[J] = C *)
    else search := 0    (* ... C not found *)
  end; (* of search *)




function b_search (C : integer) : integer;
   var U,I,L : integer;
       (* U and L are lower and upper indices *)
       (* of the array to be searched *)
       found : boolean;
   begin
     L := 1; U := N;    (* initialise the bounds of the array *)
     found := false;
     (* The array A[L]..A[U] is searched. While C is not
        found and the array is not empty the search continues *)
     while (not found) and (U >= L) do
       begin
         I := (U + L) div 2;  (* find the middle of the array *)
         if C = A[I] then found := true
         else if C > A[I] then
                 L := I + 1
              else (* C < A[I] *)
                 U := I - 1
       end;
     if found then b_search := I
     else b_search := 0
   end; (* of b_search *)
```

This new algorithm is known as 'binary search', since it repeatedly divides the array into two equal-sized sub-arrays, until either C is found or the sub-array is empty. Informally, it can be seen that each time the body of the **while** loop is executed, the size of the array is reduced by half. So, the body of the loop is executed a maximum of seven times. That is, at most seven elements of the array have to be examined. This, of course, is an immense improvement over the sequential search which may have to look up 128 elements. These two functions, although they perform the same task, are radically different in the time they need to obtain the solution.

Let us look at a second example. We wish to generate a sequence of numbers a_n (for a given $n, n \geqslant 0$):

$$0 \quad 1 \quad 1 \quad 2 \quad 3 \quad 5 \quad 8 \quad 13 \quad 21 \quad \dots$$

where

$$a_0 = 0$$
$$a_1 = 1$$
$$a_n = a_{n-1} + a_{n-2} \text{ for } n > 1.$$

Again, we can develop different solutions to get the specified result. A simple approach is to write a function A to compute a_n and then a procedure seq1 would call A repeatedly to produce a_0 to a_n.

```
procedure seq1(n : integer);
  var i : integer;
  function A(n : integer) : integer;
    begin
      if n = 0 then
            A := 0
      else if n = 1 then
                A := 1
            else A := A(n-1) + A(n-2)
    end;  (* of A *)
  begin
    if n < 0 then error
    else for i := 0 to n do
            writeln(A(i))
  end;  (* of seq1 *)
```

A second approach can use the fact that to compute a_n, we need values a_{n-1} and a_{n-2} (that is, its previous two elements). Since we generate the sequence from a_0 to a_n, we can compute a_i very simply by maintaining the last two elements of the sequence (a_{i-2} and a_{i-1}) and summing them.

This second procedure seq2, maintains the last two elements of the sequence (that is, L_0 and L_1 corresponding to a_i and a_{i-1} (for $i > 1$) in the **for** loop, and thus computes the next element of the sequence. It can readily be seen that seq1 generates a_k (for some k) many times in the recursive sequence, whereas seq2 generates a_k only once. Figure 1.1 shows the pattern of procedure calls made to the procedure A when A(5) is required.

To compute A(5): A(0) is called three times, A(1) is called five times, A(2) is called three times, and so on. Also bear in mind that A(0) and A(1) are called many more times to compute A(4), A(3) etc. Obviously, these computations are superfluous and incur great overheads on the total time needed to generate the sequence. In seq2, however, these over-computations are avoided by generating each element of the sequence only once. The mathematical analysis of these two

```
procedure seq2(n : integer);
  var L₀, temp, L₁, i : integer;
  (* L₀ and L₁ are the last two elements of the sequence at any
      instance, i.e. aᵢ₋₂ and aᵢ₋₁ *)
  begin
    (* initialise the last two elements as a₀ and a₁ *)
    if n < 0 then error
    else
      begin
        L₀ := 0; L₁ := 1;
        for i := 0 to n do
            begin
                writeln(L₀);  (* element aᵢ *)
                temp := L₁;
                L₁ := L₀ + L₁;
                L₀ := temp
            end
    end
end;  (* of seq2 *)
```

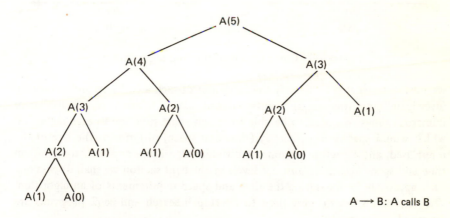

Figure 1.1 Calls to procedure A of the seq1 program.

procedures will be discussed later (in section 1.5). However, experimental measurement of the time required by these procedures shows the drastic improvement which seq2 achieves over seq1. Figure 1.2 shows experimental testing for the four procedures discussed so far.

As N, the size of the array, gets larger in figure 1.2a, the b_search function is much faster than the search function. Similarly, figure 1.2b shows that, compared with the seq2 procedure, seq1 takes a long time to execute for large values of *n*. Unfortunately, such experimental comparisons of algorithms are not very reliable in practice, because of variations in the speed of different computers, different compilers etc. Furthermore, it may be very impractical to test an algorithm for different sizes of input and all other possible cases. Ideally, we

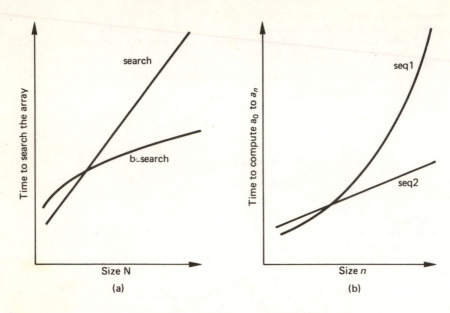

Figure 1.2 Comparisons of the four algorithms.

would like to be able to identify the behaviour of such algorithms and rigorously investigate their time requirements without having actually to execute them numerous times. We need to be able to attach some function to each algorithm which would specify its efficiency (time and space) in terms of the size of the input used, and therefore to compare different algorithms and their trade-offs in time and space at a more abstract level. In the next section we shall discuss one such approach for measuring the time and space requirements of an algorithm. The techniques which were used to develop b_search and seq2 programs are discussed in chapter 11.

1.2   Time and Space Complexity of Algorithms

The actual time and space required by a program when measured in seconds of computer time and number of bytes of computer memory will be heavily dependent on the particular computer being used. Computers with architectural differences, different instruction sets and having varying speeds of operational hardware will show different time and space measurements for a given algorithm expressed in a high-level language such as Pascal. These variations are made worse by the fact that different compilers might not translate a given Pascal program into the same machine code program (or an intermediary language such as p-code). Thus we must look for a slightly more simplified and abstract notion of computing time and space which is more or less independent of any real computer. Such an abstract model can then be used as the basis for comparing different algorithms. This approach is particularly justified since, in practice, we are often interested in the *rate* at which the time (space) requirement of a program increases as the size of the input increases. These ideas are fully discussed in section 1.4. For now, let us begin by introducing a terminology regarding the time and space complexity of an algorithm.

n − size: measure of the quantity of input data for an algorithm. For instance, in the search problem, N, the number of elements in the array, is the size of the problem whereas, in the second example the magnitude of the parameter, n, should be taken as the size. When deciding on the size of the input, we must ensure that the behaviour of the algorithm will depend on it. Indeed, to be precise, we should take as the size of the input, the total number of bits or bytes (unit of store) which are needed to represent it. Therefore, the size of the input for the search problem is actually $(\log a_1 + \log a_2 + \ldots + \log a_n)$ bits where the a_i values are the elements of the array (logs are to base 2). Similarly, in the sequence problem, the size of the input is actually $\log n$. However, in most real applications, the size of numbers is bounded by the word-length of the computer being used and thus a word of memory is taken as the unit of size. This is why in the search problem we take N as the size of the input. For the second example, however, it is not satisfactory to take the size of the input as one number since the behaviour of the algorithm for the problem is not a function of the upper bound for the argument n, but is a function of the actual number of bits in the binary representation of n. This is, of course, $(\log n)$ bits. In this case, however, we may assume that a monadic number representation is used where number n is represented as a sequence of n 'ones'. Naturally, the actual size of the input would then be n. (Chapter 12 discusses monadic and binary representations and their effects on the analysis of algorithms, see page 340.)

T(n) − *time complexity:* time needed by an algorithm to complete execution as a function of size of input n.

S(n) − *space complexity:* space (memory) needed by an algorithm to complete execution as a function of size of input n.

Because of similarities between time and space, we shall restrict ourselves to time complexity for the moment. For many algorithms the actual time and space complexities are dependent on a particular input data of size n as well as the size of data n. For instance, the search function is not only dependent on N, the size of the array, but also on the search value C. If, for instance, A[1] = C, the function search would be approximately 128 times faster than if A[128] = C or if C were not in the array at all. For this reason, it is useful to distinguish best-case, worst-case and average-case time and space complexities.

$T_{max}(n)$ – *worst-case* time complexity: maximum over all input of size n.

$T_{min}(n)$ – *best-case* time complexity: minimum over all input of size n.

$T_{avg}(n)$ – *average-case* time complexity: average over all input of size n. Usually, the assumption is made that all input sequences are equally likely. For instance, for the search problem it is assumed that C may be in any location of the array with equal probability.

The space complexity notation is similar. As we discussed before, we would like to choose an abstract notion of time and relate this to high-level languages and specific algorithms. For instance, for the search algorithm of section 1.1 we could choose the numbers of test, assignment and addition operations required as a measure of the time requirement. This is particularly suitable since a computer has a constant time for each of these operations. Now, we can derive the time complexity of the search function. For simplicity, let us assume that C indeed exists in the array (that is, for some i, $1 \leqslant i \leqslant 128$, $A[i]$ = C). If we use the notations 'a' for assignment, 't' for test and 's' for addition (sum) operation times, then for the function search:

$$T_{min}(n) = 1a + 2t + 1t + 1a = \underline{2a + 3t}$$

This occurs where the first element of the array contains C. The first assignment is for initialising J, two tests are for the **while** loop ($A[J] < C, J < N$), one test is for $A[J]$ = C and finally one assignment to search is needed. The worst case occurs when $A[N]$ = C. Thus

$$T_{max} = 1a + (n-1)a + (n-1)s + (2n)t + 1t + 1a$$
$$= \underline{(n+1)a + (2n+1)t + (n-1)s}$$

For the average case, we need to assume that all sequences of n integers in A are equally probable; that is, the probability that the ith element is C is $1/n$, or $P(A[i] = C) = 1/n$. If $A[i]$ = C, then T_i, the time needed would be

$$T_i = (i+1)a + (2i+1)t + (i-1)s$$

and

$$T_{avg}(n) = \sum_{i=1}^{n} T_i \times P(A[i] = C) = \sum_{i=1}^{n} T_i \times 1/n = 1/n \sum_{i=1}^{n} T_i$$

$$= 1/n \sum_{i=1}^{n} (i+1)a + (2i+1)t + (i-1)s$$

$$= 1/n \left(n + \frac{n(n+1)}{2} \right) a + 1/n \left(n + 2 \frac{n(n+1)}{2} \right) t + 1/n \left(\frac{n(n+1)}{2} - n \right) s$$

$$= \left(\frac{n+3}{2} \right) a + (n+2)t + \left(\frac{n-1}{2} \right) s$$

Note that $T_{avg} = (T_{max} + T_{min})/2$; that is, the average-case analysis is the average of the best-case and worst-case analyses. However, this result cannot be generalised for all algorithms (for instance, see the tree-searching algorithm in chapter 4).

As an abstract measure for space, we can take one integer to be the unit of space and thus $S(n) = 1$ for the search function. This is because only one variable, J, is used in the function.

The time and space complexities of the b_search function can be found similarly. The crucial point is that, in the worst case, the body of the **while** loop is executed $\log_2 n$ times. This is because, initially an array of size n is considered. In the second execution of the loop an array of size $n/2$ is considered and so on:

$$n, n/2, n/4, \ldots, 1$$

This, therefore, implies that the **while** loop is executed a maximum of $\log_2 n$ times. It can be shown that:

$$T_{max}(n) \approx 3(\log_2 n)t + 2(\log_2 n)a + 2(\log_2 n)s + (\log_2 n)m + (\log_2 n)d$$

and

$$S(n) = 4$$

where 'm' and 'd' stand for subtraction (minus) and division respectively. A different way to analyse this function will be discussed in section 1.5. In the following sections we shall show that the growth rate of time complexities is usually more important than absolute time measurements, and we develop a notation to express such growth rates.

1.3 Asymptotic Time Complexity — Big O and Big Ω Notations

As can be seen from the formulae of time complexities of the function search, specifying time complexities can become very complex as we consider other types of operations (for example, logical operations, array access etc.). Usually, in practice, we only require the behaviour of an algorithm when the size of the input gets very large. Furthermore, we are more interested in the rate at which time and space complexities grow rather than in absolute complexities. For instance, in the search problem, we are more interested in the fact that, when the size of the problem *doubles*, the time complexity of the function search approximately *doubles* whereas the time complexity of the function b_search only increases by a *constant* amount of time. In such circumstances we can legitimately assume that simple operations such as assignments, tests etc. take the same amount of time as *one* unit of time and also, for large n, small co-efficients and constants can be eliminated. For instance, we can assert

search: $T_{max}(n) \approx n$ units of time and
b_search: $T_{max}(n) \approx \log_2 n$ units of time

These gross approximations indeed yield the general behaviour of an algorithm for increasing large values of n (size) rather than exact measures of the time requirement. They indicate the growth rate of time complexities. When n, the size of input, gets very large, such simplifications yield what is known as the *asymptotic* complexities.

Big O and Big Ω Notations

We say $T(n) = O(f(n))$ if there exist constants C and n_0 such that $T(n) \leqslant Cf(n)$ for all $n \geqslant n_0$. In other words, $T(n)$ is bounded above by the function $f(n)$ for large n. $T(n)$ is said to be the *order of $f(n)$*. For example

$$3n = O(n): 3n \leqslant 4n \qquad\qquad \text{for all } n \geqslant 1$$
$$n + 1024 = O(n): n + 1024 \leqslant 1025n \qquad \text{for all } n \geqslant 1$$
$$3n^2 + 5n + 3 = O(n^2): 3n^2 + 5n + 3 \leqslant 11n^2 \text{ for all } n \geqslant 1$$
$$n^2 = O(n^3): n^2 \leqslant n^3 \qquad\qquad \text{for all } n \geqslant 1$$
$$\text{But } n^3 \neq O(n^2)$$

For the search function above, we can find its asymptotic behaviour as

$$T_{min}(n) = 2a + 3t$$
$$\leqslant 2X + 3X \text{ where } X = \max(a, t)$$
$$\leqslant 5X$$
$$\leqslant C \times 1 \qquad \text{where } C = 5X \text{ (a constant) for all } n$$

Therefore, according to the definition of Big O:

$$T_{min}(n) = O(1)$$

For large n, the difference between assignment and test times can therefore be discarded. Similarly

$$T_{avg}(n) = \left(\frac{n+3}{2}\right)a + (n+2)t + \left(\frac{n-1}{2}\right)s$$

$$= O(n)a + O(n)t + O(n)s$$
$$= O(n)$$

and

$$T_{max}(n) = (n+1)a + (2n+1)t + (n-1)s$$
$$= O(n)a + O(n)t + O(n)s$$
$$= O(n)$$

$T_{min}(n) = O(1)$ is said to be of *constant* time and
$T_{avg}(n) = O(n)$ is said to be of *linear* time.

To simplify the derivation of asymptotic time complexities of algorithms, certain arithmetic operations can be performed on them. A list of these is given below. For brevity, f and g are used instead of f(n) and g(n) in some places.

- $O(f) + O(g) = O(\max(f, g))$
- $O(f) + O(g) = O(f+g)$
- $O(f) . O(g) = O(f . g)$
- If $g(n) \leqslant f(n)$ for all $n \geqslant n_0$ (for a given n_0)
    then $O(f) + O(g) = O(f)$
- $O(cf(n)) = O(f(n))$ c is a constant
- $f(n) \qquad = O(f(n))$

Note that the Big O is a mechanism for finding an upper bound for the growth rate of time complexity of an algorithm and *not* a least upper bound. This is why the first two formulae above give different upper bounds for $O(f) + O(g)$. These are used in different circumstances depending on what upper bound is to be found (see also the end of this chapter).

The proofs of these rules can be derived from the definition of Big O. For instance

Lemma: If $T(n) = O(f(n))$ and $G(n) = O(g(n))$,
    then $T(n) + G(n) = O(\max(f(n), g(n)))$.

Proof

$T(n) = O(f(n)) \Rightarrow$ there exists n_1 and c_1 such that $T(n) \leqslant c_1 f(n)$ for all $n \geqslant n_1$.
$G(n) = O(g(n)) \Rightarrow$ there exists n_2 and c_2 such that $G(n) \leqslant c_2 g(n)$ for all $n \geqslant n_2$.
Let us assume

$$n_3 = \max(n_1, n_2)$$
$$c_3 = \max(c_1, c_2) \text{ and}$$
$$h(n) = \max(f(n), g(n)) \text{ for all } n$$

Since $n_3 \geqslant n_1$: $T(n) \leqslant c_1 f(n)$ for all $n \geqslant n_3$.
Since $c_3 \geqslant c_1$: $T(n) \leqslant c_3 f(n)$ for all $n \geqslant n_3$.
Since $h(n) \geqslant f(n)$: $T(n) \leqslant c_3 h(n)$ for all $n \geqslant n_3$.
Similarly: $G(n) \leqslant c_3 h(n)$ for all $n \geqslant n_3$.
Therefore: $T(n) + G(n) \leqslant c_3(h(n) + h(n))$
$$\leqslant 2c_3 h(n).$$

Thus: $T(n) + G(n) = O(f(n)) + O(g(n)) = O(h(n))$ where $h(n) = \max(f(n), g(n))$ and the proof is complete. The other rules can be proved in a similar manner.

As an example of using these rules, let us consider the following double-loop Pascal statement:

```
for I := 1 to n do
    for J := 1 to I do
        begin
            S1;  S2;  S3;  S4
        end;
```

Each S_I is a simple assignment. The asymptotic time complexity of this statement can now be determined using the above rules.

Time for each S_I: $O(1)$

Time for **begin** S1; S2; S3; S4 **end**;: $O(1) + O(1) + O(1) + O(1) = O(1)$

Time for the inner loop: $I.O(1) = O(I).O(1) = O(1.I) = O(I)$

Time for the outer loop $= \sum_{I=1}^{n} O(I)$

Using the second rule $= O\left(\sum_{I=1}^{n} I\right) = O\left(\frac{n(n+1)}{2}\right)$

$$= O(1/2n^2 + 1/2n)$$
$$= O(1/2n^2)$$
$$= O(n^2)$$

Therefore, the whole statement has asymptotic time complexity $O(n^2)$. The Big O notation enables us to define an upper bound for the growth rate of time complexity of an algorithm. To find a lower bound for time complexity, we define Big Ω as:

$T(n) = \Omega (f(n))$ if there exists a constant C such that $f(n) \geqslant Cf(n)$
for *infinitely many* n

The reason that Big Ω is defined for infinitely many n and not for all n greater than a given number is that many algorithms are very fast on some sizes of input but not all. For example

$$T(n) = \begin{cases} 2n & n \text{ is odd} \\ 3n^2 - 10 & n \text{ is even} \end{cases}$$

$T(n) \geqslant n^2 \quad n = 4, 6, 8, \ldots$

Therefore $T(n) = \Omega (n^2)$.

1.4 Importance of Growth Rate of Time Complexities

One might reason that, since computing times are steadily decreasing, we do not have to worry so much about developing efficient algorithms to solve problems. We can simply run our programs on a more advanced computer. Unfortunately, this does not work in practice for two reasons. Firstly, computers tend to be presented with more and more sophisticated instances of a problem and it may not be practical to keep on moving to more advanced machines (faster). Secondly, and more importantly, the growth rate of the time complexity of an algorithm is the limiting factor which determines if larger problems can be solved in a reasonable amount of time. To clarify this point, let us consider six algorithms A1, A2, . . ., A6 and observe their behaviour as they are run on two computers; one 60 times faster than the other one. Assuming that the unit of time is one millisecond, the maximum problem sizes $S_1, S_2, . . ., S_6$ are given for the six algorithms which can be processed in one minute (6×10^4 milliseconds) of computing time (table 1.1).

Table 1.1   Comparison of speed-up factors for six algorithms

Algorithm	Time complexity	Max. problem size in 1 minute	Max. problem size in 1 hour	Max. problem size in 1 minute after speed up
.A1	n	$S_1 = 6 \times 10^4$	$S_1' = 3.6 \times 10^6$	$S_1' = 60S_1$
A2	$n \log n$	$S_2 \approx 4893$	$S_2' \approx 2 \times 10^5$	$S_1' \approx 60S_2$
				(for large S_2)
A3	n^2	$S_3 \approx 244$	$S_3' \approx 1897$	$S_3' \approx 8S_3$
A4	n^3	$S_4 \approx 39$	$S_4' \approx 153$	$S_4' \approx 4S_4$
A5	2^n	$S_5 \approx 15$	$S_5' \approx 21$	$S_5' \approx S_5 + 6$
A6	$n!$	$S_6 \approx 8$	$S_6' \approx 10$	$S_6' \approx S_6 + \epsilon$

$$\text{for } \epsilon \approx \frac{\log_e 60}{\log_e S_6 - 1}$$

Also the maximum problem sizes are given in one hour (36×10^5 ms) of computation: $S_1', S_2', . . ., S_6'$. Now, if the speed of the computer is increased 60 times, S_i' represents the maximum problem size executable in one minute for the problem Ai. It is obvious that a linear algorithm such as A1 yields a factor of 60 increase, a quadratic algorithm such as A3 yields a factor of about 8 while algorithms A5 and A6 yield only a small constant increase. Such algorithms limit the size of problems which can be solved in a reasonable amount of time. For example, algorithm A5 needs almost 6 centuries of computing time to solve a problem of size 50 (before speed up of 60 times)! The last column in table 1.1 shows the relationship between S_i' and S_i which is an indication of how much bigger problems can be solved in the same amount of time after speed up of 60 times. These relationships can be found by equating $60 \times f(S) = f(S')$ for an algorithm with time complexity $f(n)$. For instance, for algorithm A4, $60 \times S_4^3 = S_4'^3$. Thus $S_4' = 60^{1/3} \times S_4$, that is, $S_4' \approx 4S_4$.

When faced with solving more complicated examples of a problem, polynomial and logarithmic algorithms such as A1 to A4 give *multiplicative* size increases as the speed of the computer is increased or more computer time is made available. This can be contrasted with algorithms such as A5 and A6 which only give *additive* size increases.

So it can be seen that asymptotically efficient algorithms are those which have polynomial or logarithmic time complexities. These are known as *polynomial* algorithms, since they can be bounded by a polynomial; that is, there exists a polynomial $P(n)$ such that $f(n) \leqslant CP(n)$ for all $n \geqslant n_0$ where C and n_0 are two constants and $f(n)$ is the time complexity of the algorithm. For instance, A2 is a polynomial algorithm since $\log n \leqslant n$ for all $n \geqslant 1$. Algorithms which are not polynomial are called *exponential*. Algorithms A5 and A6 are exponential. Almost all the algorithms described in this book are polynomial algorithms. Certain classes of problems may only have exponential algorithms. Such problems are classified as very *hard* problems. These and some techniques for deriving good algorithms for them are discussed in chapter 12. The spectrum of algorithm complexity can be illustrated as follows:

$$1 < \log \log n < \log n < n < n \log n < n^2 < n^3 < \ldots < 2^n < n!$$

Note that although an exponential algorithm is asymptotically worse than a polynomial one, it may however, for small values of n, be a better algorithm. For instance, $2^n \leqslant 100n^2$ for $n = 1, 2, \ldots, 13, 14$. So for small n, it may be best to use an exponential algorithm. Also, if an algorithm is to be executed only a few times, the efforts to find an efficient algorithm may not be justified and hence a slower algorithm may be preferred. Finally, it should be borne in mind that asymptotic complexities are only a gross measure of cost and for more exact comparisons the coefficients and constants in the complexity formulae should be found. For instance, although $1000n \log n$ and $1/10n \log n$ are both $O(n \log n)$, the latter is far superior to the former one.

1.5   Rules for Deriving Time Complexities of Pascal Programs

In general, average-case and best-case time complexities are very difficult to derive. This is partly because information regarding the probability of each permutation of input data of size n may not be known, or cannot be determined. Furthermore, for many algorithms, determination of average-case and best-case complexities are mathematically very difficult. We shall meet some of these in later chapters. Because of this, it is usually only reasonable to look for worst-case time complexities of algorithms. Here we shall discuss some general rules regarding determination of worst-case complexities.

In finding the *worst-case* time complexity of a Pascal program the following guidelines can be used:

1. Assignment, comparisons, arithmetic operations, read, write (a constant or a single variable) are of $O(1)$.
2. **If C then S$_1$**
   else S$_2$;

needs $T_c + Max(T_{s_1}, T_{s_2})$ where T_c, T_{s_1} and T_{s_2} are time complexities of C, S_1 and S_2.

3. **For** loops. The number of times the loop is executed must be multiplied by the time complexity of the body of the loop.

4. **while** C **do** S; and **repeat** S **until** C;

For both these loop constructs we should find the number of times the loop will be executed and multiply this by the complexities of C and S. For instance, the loop of b_search function is executed $\log_2 n$ times. Thus, its time complexity is $\log_2 n (O(1)) = O(\log_2 n)$. This simple analysis may not work for all loops. In these cases other properties of the program should be used to infer the time complexity. As an example, let us look at the following hypothetical program fragment:

```
size := m;
i := 1;
while i < n do
  begin
    i := i + 1;
    statement₁; ................ O(n)
    if size > 0 then
      begin
        Choose a number in the range 1..size and assign it to s;
        size := size - s;
        for j := 1 to s do
          statement₂; ......... O(n)
      end
  end;
```

The **while** loop above will be executed $n-1$ times. This is because i is initialised to 1 and is incremented by 1 each time the body of the **while** loop is executed. Thus statement₁ will be executed $n-1$ times. We cannot find the number of times the **for** loop (inside the **while** loop) will be executed as a function of i. However, if s_i is the value of s inside the **while** loop for a given value of i, then the total number of times statement₂ will be executed is $\sum_i s_i$. The sum of the s_i values will be $\sum_i s_i = m$, since size is initially set to m and in each iteration of the **while** loop s_i is subtracted from size. Thus the total time due to statement₂ will be $m \times O(n)$. The total time due to statement₁ is $n \times O(n)$. The time complexity of the entire while loop is

$$n \times O(n) \quad + m \times O(n) \quad + \quad O(1) \quad = O(n^2 + mn)$$

statement₁　　statement₂　　tests, assignments etc.

We shall see examples of such loops in the rest of the book (for instance, see Dijkstra's algorithm of chapter 7).

5. **Case A of** $a_1 : S_1 ; a_2 : S_2 ; \ldots ; a_n : S_n$ **end**;
   The time complexity is $\mathrm{Max}(T_{s_1}, T_{s_2}, \ldots, T_{s_n})$.
6. Accessing individual elements of an array and selecting individual fields of a record take $O(1)$ time.
7. Procedures and functions

The same rules described above can be used to determine the complexity of procedures and functions. Care must be taken that the parameters and global data structures are taken into account. When procedure and function calls are made, the time to transfer the control to the calling routine can be assumed to be constant $(O(1))$, and can generally be ignored.

However, for recursive procedures and functions, the problem is slightly more complex. The usual technique for deriving complexities of recursive routines is to use *recurrence equations* or *relations*.

A recurrence equation is an equation which defines the time complexity of a recursive routine in terms of complexity of smaller versions of the same problem.

As an example, let us look at the following function which is in fact a recursive formulation of the b_search function discussed in section 1.1.

```
function b_search(C,L,U : integer) : integer;
   (* search A[L] .. A[U] to find C and return its index *)
   var    index, element : integer;                                  UNITS
   begin
      if (U < L) then                                                  [3]
          b_search := 0           (* the sequence is empty *)         [1]
                                   (* search unsuccessful *)
      else begin
             index := (L + U) div 2; (* the middle of the array *)    [3]
             element := A[index];                                     [2]
             if element = C then                                      [1]
                 b_search := index  (* search successful *)           [1]
             else if element > C then                                 [1]
                    b_search := b_search (C, L, index - 1)    [3 + T(n/2)]
                  else
                    b_search := b_search (C, index + 1, U)    [3 + T(n/2)]
      end
   end; (* of b_search *)
```

Now, $T(n)$, the worst-case time complexity of this function can be defined in terms of the length of the sequence, $U - L + 1 \equiv n$. The number of computational units required for each step in the program is given above. Based on these we can derive

$$T(n) = \begin{cases} 1 + 1 & n = 0 \\ 1 + 3 + 2 + 1 + 1 + 3 + T(0) & n = 1 \\ 1 + 3 + 2 + 1 + 1 + 3 + T(n/2) & n > 1 \end{cases}$$

The assumption is made that C is not in the array (and thus the complexity is the worst case). For $n = 0$, one test $(U < L)$ and one assignment are needed. For $n = 1$, one test $(U < L)$, two arithmetic and an assignment (index := (L+U) **div** 2), one array access and an assignment (element := A[index]), one test (element = C) and finally a recursive call are needed, the latter requiring one subtraction, one assignment and a control transfer to the procedure. This assumes the worst case when C is not in the array. When $n > 1$, a similar expression can be formed. Thus

$$T(n) = \begin{cases} 2 & n = 0 \\ 13 & n = 1 \\ 11 + T(n/2) & n > 1 \end{cases}$$

To solve this recurrence relation we proceed as follows:

$$\begin{aligned} T(n) &= 11 + T(n/2) \\ &= 11 + 11 + T(n/4) = 2 \times 11 + T\left(\frac{n}{2^2}\right) \\ &= 3 \times 11 + T\left(\frac{n}{2^3}\right) \\ &\quad\vdots \\ &= k \times 11 + T\left(\frac{n}{2^k}\right) \end{aligned}$$

So, when $n/2^k = 1$, we can solve the recurrence relation by substituting the value of $T(1)$ in the above formula.

$$\frac{n}{2^k} = 1 \Rightarrow n = 2^k \Rightarrow k = \log_2 n$$

Therefore

$$T(n) = 11 \log_2 n + 13$$

and

$$\underline{T(n) = O(\log_2 n)}$$

Recurrence relations such as this one occur very frequently in the analysis of computer algorithms. For the purpose of the algorithms in the remainder of this book, we can generalise this result and give the following formulae for solving similar recurrence relations:

$$a, b, c \geqslant 0$$

$$(1) \quad T(n) = \begin{cases} b & n=1 \\ aT\left(\frac{n}{c}\right) + bn & n > 1 \end{cases}$$

$$\text{Then } T(n) = \begin{cases} O(n) & a < c \\ O(n \log_c n) & a = c \\ O(n \log_c n) & a > c \end{cases}$$

$$(2) \quad T(n) = \begin{cases} b & n = 1 \\ aT\left(\dfrac{n}{c}\right) + bn^2 & n > 1 \end{cases}$$

$$\text{Then } T(n) = \begin{cases} O(n^2) & a < c^2 \\ O(n^2 \log_c n) & a = c^2 \\ O(n \log_c n) & a > c^2 \end{cases}$$

Using techniques similar to the one used in the b_search function, these formulae are easy to verify (see exercise **1.17**).

Let us return to the seq1 function and examine its time complexity. The function A is recursive. Let us first look at the asymptotic complexity of A. If $F(n)$ denotes the time complexity of function A, in a similar way to the b_search above, we can formulate $F(n)$ as

$$F(n) = \begin{cases} 1 + 1 & n = 0 \\ 1 + 1 + 1 & n = 1 \\ 1 + 1 + F(n-1) + F(n-2) + 2 + 1 + 3 & n > 1 \end{cases}$$

When $n = 0$, one test and one assignment are necessary. For $n = 1$, two tests and one assignment are needed and finally for $n > 1$, two tests, two calls to the same function of sizes $n-1$ and $n-2$ (that is, $F(n-1)$ and $F(n-2)$) and one addition, two subtractions and one assignment, are needed. Therefore

$$F(n) = \begin{cases} 2 & n = 0 \\ 3 & n = 1 \\ 8 + F(n-1) + F(n-2) & n > 1 \end{cases}$$

Now the time for the entire procedure seq1 can be determined. Let $T(n)$ be the time complexity of seq1. Seq1 only calls function A $n+1$ times. Thus

$$T(n) = F(0) + F(1) + F(2) + \ldots + F(n) + 1$$

$$T(n) = 1 + \sum_{i=0}^{n} F(i)$$

To find $T(n)$, first a closed formula should be found for $F(n)$ (in terms of n) and then substituted in $T(n)$. A close examination of the recurrence relation F shows that the techniques used for the b_search function does not yield any results. Instead, for $F(n)$ we have to use some approximation techniques. To begin with let us introduce the recurrence relation $R(n)$ as

$$R(n) = \begin{cases} 1 & n = 0 \\ 1 & n = 1 \\ R(n-1) + R(n-2) & n > 1 \end{cases}$$

It is clear that $F(n) > R(n)$ for all n. Now we should try to find a closed formula for $R(n)$. We can find such a formula using the approximation

$R(n) = k\,x^n$ for some constants k and x

If we substitute this value in the above recurrence relation we get

$R(n) = R(n-1) + R(n-2)$
$k\,x^n = k\,x^{n-1} + k\,x^{n-2}$

Dividing both sides by x^{n-2} :

$x^2 = x + 1$
$x^2 - x - 1 = 0$

$$x = \frac{1 \pm \sqrt{(1+4)}}{2}$$

$$x = \frac{1 \pm \sqrt{5}}{2}$$

$x = (1 + 2.223/2) = 1.618 \approx 1.62$

Thus: $R(n) = k\,(1.62)^n$ for some k.
Now $F(n) > R(n)$, therefore

$F(n) > k\,(1.62)^n$ for some constant k

According to the definition of Ω we can assert

$F(n) = \Omega\,(1.62^n)$

This indicates that algorithm A requires at least an exponential amount of time. To calculate $T(n)$:

$$T(n) = 1 + \sum_{i=0}^{n} F(i)$$

$T(n) > 1 + x + x^2 + x^3 + \ldots + x^n$ where $x = 1.62$

$$> \frac{x^{n+1} - 1}{x - 1}$$

$$> 1.618\,(x^{n+1} - 1)$$

Therefore we conclude that $T(n) = \Omega\,(1.62^{n+1})$. Again this suggests that the procedure seq1 is an exponential algorithm. However, the procedure seq2 can be easily shown to be $O(n)$. Similarly, the function search of section 1.1 can be shown to be $O(n)$. These growth rates clearly match the experimental results of figure 1.2. As n gets larger, $O(n)$ gets larger much faster than does the $O(\log n)$ algorithm. Equally, $\Omega(1.62^{n+1})$ behaves similarly when compared with the $O(n)$ algorithm of the seq2 program.

Before ending this chapter we should mention that when evaluating space complexity of algorithms, similar rules to the above can be followed. However, for recursive routines, extra implicit space is used for maintaining a stack. Recursive routines are implemented by storing the local information and return address on top of a stack every time a recursive call is made. The size of the stack needed is proportional to the maximum level of recursion of the recursive routine at any instant. For instance, the b_search function above uses $E(n) = O(\log_2 n)$ space for the stack in the worst case. This is because in the worst case (when C is not in the array) b-search calls itself a maximum of $E(n)$ times; since by then the size of the array $A[L]\ldots A[U]$ will be 1. A more convincing way to show this is by using a recurrence relation $E(n)$, the extra implicit storage for the stack:

$$E(n) = \begin{cases} k & n = 1 \text{ or } 0 \\ \\ k + E(n/2) & n > 1 \end{cases}$$

where k is the storage used for one invocation of b_search (a constant). When n is 0 or 1, k units of space are used for the current call; when $n > 1$, k units of space are used for the current call and $E(n/2)$ is used for the only procedure call. This recurrence formula can be solved as before, yielding $E(n) = O(\log_2 n)$. Likewise, the extra implicit space used for the seq1 procedure can be shown to be $O(n)$.

Finally, a word of warning regarding interpretation of Big O functions. Big O only indicates an *upper bound* for the time requirement of an algorithm and not a *least upper bound*. For instance, as we have seen before, an algorithm with time complexity n is $O(n^2)$. Although this algorithm is $O(n^2)$, a more restricted bound would be $O(n)$. To get a more meaningful understanding of growth rates we must show that: $T(n) = O(f(n))$ and $T(n) = \Omega(f(n))$, that is, $f(n)$ is an upper bound and a lower bound for $T(n)$. For most algorithms in this book, it turns out that the upper and lower bounds coincide. In general, however, care must be taken not to make false assumptions about Big O functions.

Exercises

1.1. What does the following procedure do?

```
(* A is an ordered array [1..n] of integer *)
procedure search (C : integer; var I : integer);
  var J : integer;
      found : boolean;
  begin
    J := 1; I := 0; found := false; (* initialise J, I and found *)
    while not found and J <= n do
          if A[J] = C then
                found := true
          else J := J + 1;
    If found then I := J
  end;
```

How does this procedure compare with the one given in section 1.1?

1.2. In the sequential search program of section 1.1, what are the best-case, worst-case and average-case complexities in terms of assignment and test operations if C *may not* be in the array? State your assumptions.

1.3. Modify the function search of section 1.1 so that it would be suitable for unordered arrays. What is the average time complexity of this new function?

1.4. In the search problem of section 1.1, assume that the array A is defined as

A: **array** [1. .n] **of** integer; *in non-decreasing order*

Are the two sequential and binary search algorithms equivalent? If not, why not?

1.5. In the sequential search program, modify the list as

A: **array** [1. .nPlus1] **of** integer;

where nPlus1 = n + 1 and the array is not in sorted order. Naturally, a slightly modified version of the function search can be used. An alternative solution would be: to search for C, first execute A[nPlus1] := C; then search sequentially for C. This loop is bound to find C. Finally if C is found in A[nPlus1], the search has failed. Implement these two programs and compare their execution times for a large sequence (n ≫ 1000). Which one is faster and why?

1.6. Consider the following two programs which find A[m], the minimum element of an unordered array A[1. .n]:

What are average-case, worst-case and best-case time complexities of these algorithms? Which one is fastest? Asymptotically?

```
A : m := 1;                          B : m := 1;
    min := A[1];                         for i := 2 to n do
    for i := 2 to n do                       if A[i] < A[m] then
        if A[i] < min then                       m := i;
        begin
            m := i;
            min := A[m]
        end;
```

1.7. An English dictionary is an ordered list of words with approximately 1100 pages. To search for a particular word using binary search we need to examine a maximum of 10 pages. Is there a better way of doing the search? Assume that the distribution of words is uniform; that is, the number of words beginning with 'A', 'B', ... and 'Z' are approximately equal. *Hint:* Divide the sequence not in the middle, but at a place more likely to be closer to the actual location (interpolation search).

1.8. Write bad and good programs to evaluate polynomials. For instance, evaluate

$$2x + 4x^2 + 5x^3 + 13x^4$$

at $x = 5$. Design the necessary data structures and write the programs. Compare their time complexities.

1.9. Design a recursive algorithm for computing x_n (assume n is a power of 2) of the sequence

$$x_0 = 0$$
$$x_1 = 1 \quad x_2 = 1$$
$$x_n = x_{n/2} + 3x_{n/4}$$

1.10. Formulate a non-recursive efficient algorithm for the sequence of exercises **1.9** by not evaluating x_k more than once (for a given k).

1.11. What is the time complexity of your solution to exercise **1.10**.

1.12. Write a program to compute e

$$e(n) = n + \frac{n^2}{2!} + \frac{n^3}{3!} + \frac{n^4}{4!} + \ldots + \frac{n^{10}}{10!}$$

What is the time complexity of your solution?

1.13. Find the asymptotic equivalent of the following functions:

$$3n^2 + 10n$$

$$\frac{1}{10} n^2 + 2^n$$

$$21 + \frac{1}{n}$$

$$\log n^3$$
$$10 \log 3^n$$

$(f(n) + 10)^2$ when $f(n) = O(n)$

1.14. Which of the following are valid?

$$O(1) = O(100)$$
$$O(1) = 100$$

1.15. Show that if a computer executes algorithm A6 of size S in T units of time, then a computer which is 60 times faster would execute the algorithm in T time for size x where

$$x = S + \frac{\log_e 60}{\log_e S - 1}$$

Hint: $n! \approx \left(\frac{n}{e}\right)^n$

1.16. Can you give guidelines, as is done in section 1.5, for finding the best-case time complexities? If not, why not?

1.17. Verify the solutions to the recurrence relations of section 1.5.

Bibliographic Notes and Further Reading

The discussion of the complexity theory in this chapter is very pragmatic. In order to reason about the behaviour of algorithms more accurately and for comparing algorithms, more precise and mathematical models of computation are needed. These include the Turing Machine, Decision Trees, the Random Access Machine etc. More on these models and their relationship to one another can be found in Aho *et al*. (1974).

The sequence problem of section 1.1 is the famous Fibonacci series. Note that in the seq1 procedure, the time complexity of the program to generate the sequence was bounded below by the sequence itself.

In chapter 12, where 'hard' problems are discussed, some issues regarding exponential algorithms are discussed and the bibliographical notes give more relevant references.

More on the interpolation search of exercise **1.7** which is of $O(\log \log n)$ can be found in Gonnet *et al*. (1980). Apart from best-case, worst-case and average-case complexities, there is also *amortised* time complexity which is a measure of executing a program M number of times. For instance, when maintaining a linear list with the search operation it is possible to have an unordered list, but every time an element is searched it is moved to the front of the list. In this example, although the worst-case complexity for one search operation remains $O(n)$, however, performing n search operations may require much less than $O(n^2)$ because of some locality of accesses to the elements of the list. This particular problem and a few algorithms are further discussed in Sleator and Tarjan (1985). Another example of this type of analysis is applied to the data type 'Component_Set' of Kruskal's algorithm in section 7.3.

Aho, A., Hopcroft, J. and Ullman, J. (1974). *The Design and Analysis of Computer Algorithms*, Addison-Wesley, Reading, Massachusetts.

Gonnet, G. H., Rogers, L. D. and George, J. A. (1980). 'An algorithmic and complexity analysis of interpolation search', *Acta Informatica*, Vol. 13, pp. 39–52.

Sleator, D. D. and Tarjan, R. E. (1985). 'Amortised efficiency of list update and paging rules', in *CACM*, Vol. 28, No. 2, pp. 202–208.

2 Abstract Data Types and Program Design

A computer program (at an assembly-language level) can be informally defined as a sequence of instructions for a computer to perform a particular task. The exact form and sequencing of the instructions for a given task depend largely on the underlying architecture of a computer (physical machine). Writing a computer program therefore requires decisions on how to represent the task to be achieved in terms of the constructs of the architecture. However, more often than not, the formulation of a task as a specification is very different from its representation in terms of the architectural-level instructions. Such a disparity between these two levels of programming implies a considerable intellectual and organisational effort to produce a correct program for a task. In most cases, however, the organisational details are numerous and they become difficult to handle. To reduce the amount of such detailed organisational activity, we can define 'idealised' architectures on top of the machine architecture, and in this way we will provide a new framework for programmers so that they can express their requirements more easily and effectively. In this chapter we shall briefly discuss the main components of a program using an 'assembly-language'-level architecture, and then describe a more idealised architecture which would accommodate higher-level constructs for certain programming elements: operation, control and data. The effect of using the new higher-level architecture is the removal of unnecessary details, which can be dealt with at a later stage in program development.

2.1 Elements of Program Design

It is assumed that the reader is familiar with at least one assembly language. An assembly language directly reflects the physical architecture of the computer under consideration. Figure 2.1 shows a simplified architecture of a computer.

The essential features of such an architecture are the particular representation of data, operations and control.

Data. The structure of data used at this level is a linear memory unit (MU) organised as a linear sequence of bytes or words, and also a set of registers (ALU) where each one can store one word of data.

Operations. Pieces of data are manipulated by performing *operations* on them. Such operations as arithmetic (addition, subtraction, division, multiplication) and logical operations (and, or, not etc.) (ALU), data transfer to and from memory (MU) or registers (RU) (instructions such as move and store), and

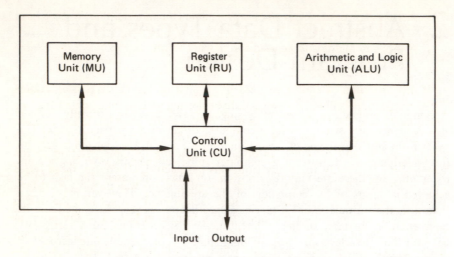

Figure 2.1 A simple computer architecture.

special operations such as shift, rotate etc. (CU) are supported at this level.
Control. In order to achieve a given task, the operations on data have to be performed in a certain sequence. This sequencing of operations is known as control. Default sequencing (';' in Pascal), different branch (goto or jump) commands, jump_to_subroutine and return_from_subroutine are among control mechanisms at the machine level (CU).

To design a program, first the data to be used in the program must be organised and allocated. Then, using the correct control constructs and operations, the data should be manipulated to produce the required results.

Using these minimal and primitive facilities the task of developing large programs is very difficult, if not impossible. Although the programs can be broken down into procedures (subroutines), however the assembly language environment lacks mechanisms for more abstract definitions of data and specification of parameters to sub-programs. Also, there is no standard method for passing parameters to sub-programs. This in turn implies that reasoning about the behaviour of programs becomes very difficult as their size and complexity grow. The major reason for this difficulty is the amount of detail (data definition, procedure definition, parameter passing etc.) which must be handled at any moment. As an example, let us consider the operation of adding an extra number to the end of a list of numbers stored in the memory. Let us denote this operation as add_to_end(n) where n should be added to the end of the list. To be able to do this, we aim to write a procedure and we should consider the following details:

- Where are the elements of this list stored in the main memory?
- Are the elements stored in consecutive locations of memory or not?
- If not, how are the elements linked together?

- How is the parameter of this procedure (the new number) to be passed to the procedure?
- Are any registers to be used and in what form?

and so on.

If this operation of adding an element to a list of numbers is part of a much larger program, it is likely that we should be dealing with so much detail that we might easily lose sight of the organisation of the entire program. In the context of a large program, all we need to know about this operation is that it adds a number at the end of a list. Information regarding its representation in the memory etc. is totally irrelevant from this perspective. Of course, these issues must be considered at some stage in order to optimise the way memory, registers, the ALU and the CU are used. But these factors need not get into the design stage. So, if we can reduce the amount of information needed at any point in time by discarding many irrelevant details, the task of developing a program becomes much easier. The principle which we will use to achieve this is the process of *abstraction*.

2.2 Abstraction Mechanisms in Program Design Process

The term 'abstraction' refers to a process which discards many details and emphasises only the 'main features' of interest at a particular 'level of concern'. Indeed, we use such abstractions in our every-day lives almost all the time. For instance, to convey the notion of 'Fred's car is red', it would be utterly impractical if we had to describe it as "The thing which has four wheels, a body consisting of an engine compartment, a boot and a passenger compartment capable of accommodating four persons; one of its headlamps does not work, . . .; it belongs to Fred who is a man, six foot tall, with dark brown hair, . . .; it is red on the outside of the body with a few stripes in silver and the wheels are also in red and . . .". In a simple conversation, the level of concern only requires the facts concerning the ownership of a car (by Fred) and its colour (which is red). Other details of the car and details of Fred are totally irrelevant at this level. It is only necessary for a person to have an abstract notion of a car, know that Fred is a person, and a person is capable of owning a car, and that cars have colours. Any more details at this level of discourse only serve the purpose of obscuring the issue and hiding the purpose of the argument. To get a better picture, imagine that in the long conversation below, every reference to Fred's car is expanded as the description above: "Fred's *car* was involved in a breakdown. All of a sudden, the engine of the *car* stopped functioning and the gear box seized. A mechanic from the Westfield Garage was called to inspect the *car* and it was diagnosed that one of the engine mounts of the *car* was broken and, as a result, the engine was displaced and had jammed the gearbox." After expanding the 'car' to its full description, it would not be easy to extract the real intention of the conversation! Also notice that this conversation is aimed at a recipient who has some knowledge of cars, so the description provides some useful information. To a recipient with no knowledge or interest in cars the conversation would probably be: "Fred's car broke down and a mechanic was called to repair it."

Just to summarise the essence of the foregoing paragraph: firstly, at a level of abstraction (concern) we wish to concentrate on essential features of the objects and not all the details and, secondly, once an object is defined we wish to refer to it by a name and assume only its prescribed features of interest. Thus, once Fred's car is described, we refer to it as 'Fred's car' and there will be no need to give a full description each time it is mentioned.

Abstraction is used in many areas of computer science to reduce the complexity of tasks to manageable levels. The example of adding to a list, quoted in the previous section, illustrates the concept of abstraction in the context of programming. We merely wish to consider the properties of a list of numbers, the operation add_to_end acting on that list and a number. A list (for example, a shopping list) is a sequence of items, where each item comes before another item and after another item, which has a beginning and an end, and to which items can be added at the end of the list. Other details such as memory allocation, optimisation of time or space etc. can all be dealt with at a lower level of description at a later stage. Furthermore, once a list is defined in this way, it can be used over and over again, and we need only consider its abstracted properties. Thus, its implementation details need only be considered once. Also, since the abstract definition of a list is separate from its implementation, the implementation can be modified to achieve more speed, to run on a different computer or to use a different programming language etc. without affecting any programs which utilise such an abstract list.

Now that we have a feel for the notion of abstraction, we can start looking for more abstract descriptions of programming elements. In high-level programming languages such as Pascal, abstraction is used to define more meaningful and succinct constructs of data, operation and control by hiding some of the unnecessary details of programming. Let us now look at each construct in turn.

Procedures and Functions

Procedures and functions are abstractions of operations. Each operation is described by its name, a few parameters and what it does. Sometimes, as a result of an operation, a piece of data is returned. For instance, 'add R1,R2,R3' means 'add registers 1 and 2 and store the result in register 3'. Using procedures and functions, the basic repertoire of machine instructions can be extended. Each procedure or function defines an operation by providing: its name; a description of its parameters; an optional value that it may return; and a specification of the relation between the parameters, the returned value and any other data values. When using such operations there is no need to refer to the body of the procedure or function which contains detailed information about how it actually produces the result. Such an abstraction has many advantages. Firstly, by defining suitable procedures and functions which reflect the structure of the problem to be solved, the task of developing the program is made easier by using these procedures and functions as the basic operations and concentrating on the overall structure of the program rather than details of each operation (compare this with the description of the break-down of Fred's car). The techniques of structured programming and stepwise refinement use these abstractions to break down a task into sub-tasks, and so on until each sub-task is simple enough to be implementable directly in the language being used (see the example later). Another advantage is that if the task to be solved by a procedure or a function is specified independently of how it may be implemented then, if at a later date its implementation needs to be modified or changed, all the changes can be done in only one place in the program without affecting the rest of the program. It is assumed that such changes are local to the body of the procedure or function and have no effect on the overall function of that routine. This is to say, the specification of the task has not changed — only the way it is achieved has changed. For example, in the search problem of section 1.1, the specification for the function was to find index i such that $A[i]$ = C, and if there was no such i, a zero was to be returned. The functions search and b_search both performed the same task but using different methods. They were two implementations for the same task.

Naturally, this is only possible if the procedures and functions are selected in such a way that each one represents a *well-specified* sub-task.

Abstract Control Constructs

As you are familiar with the highly structured control constructs of Pascal (or a similar high-level language) it is easy to see how they remove the burden of having to define explicit labels and different branch commands as in an assembly language. This is specially true as the conditions for transferring control become more complex by using logical operators and so on. The usual abstract control structures found in high-level languages are listed at the top of page 30.

An unconditional GOTO statement is also provided in most languages. However, because of its unstructured nature, it must be used very rarely. Its use is only justified when the control needs to be moved to the end of a procedure, a function or the end of the program.

The details of implementation of these abstract control constructs in terms of the lower-level machine control structures are left to the compiler of the language being used. Thus the programmer can concentrate on the essential and more important tasks of designing the program and verifying that it is correct.

```
. default sequencing ';'.
. conditional   :   if condition then statement₁ else statement₂;
. selection     :   case expression of
                            value₁:  statement₁;
                            value₂:  statement₂;
                               .
                               .
                               .
                            valueₙ:  statementₙ
                        end;
. loop          :   while condition do statement;
                    repeat statement until condition;
                    for fixed no of times do statement;
. procedure and function call (These can be recursive).
```

Data Abstraction and Data Hierarchies

At assembly-language level, the data items used are primarily bytes and words. In most assembly languages some abstraction is provided by interpreting certain words (or groups of words) as integers, reals or sequences of characters.

Corresponding operations such as arithmetic for numbers are therefore provided. In a high-level language these data items are known as *data types*. Generally, a data type is a set of *permissible* objects with a number of operations defined on these objects. Each operation is primarily a function which, when applied to an object of a given type (or several objects), returns an object of a specified type. For instance, the arithmetic operation '+' is a function

'+': integer * integer → integer

which when given two integers returns another integer. A special operator, ':=', (that is, the assignment operation) is a function with two arguments of the same type which returns no value but changes the state of the program by copying the second argument into the first argument. For example, after '$x := y;$' is executed, x will have the same value as y. Note that the first argument must be a 'variable' capable of containing a value and the second argument is either a value or a variable.

High-level languages provide further abstractions of data by grouping a number of data objects in a meaningful manner. These new objects are known as *data structures*. Data structures in Pascal include **arrays**, **records** and a restricted form of **sets**. **Files** are another data structure in Pascal which are abstractions of external storage such as magnetic discs etc. The operations on data structures can be summarised as

array: selection operation [ ] to select a particular element of an array
record: selection operation '.' to specify a named field of a record
set: operations 'in' , '+' , '−' and set creation [. . .]
file   rewrite, reset, put, get, read and write operations

Furthermore the assignment operation is defined for all these types except for files. In a similar way to abstract control constructs, these data structures are implemented efficiently by a compiler and the programmer is relieved from all the unnecessary details of their implementation.

Although this might seem to be the extent of data abstraction that a high-level language such as Pascal can provide, this is not actually so. To solve many problems the data can be described more naturally in terms of sets, lists, graphs, trees etc. than in terms of the arrays, records and data types of Pascal. Such abstract data items are termed *abstract data types*. An abstract data type is usually a mathematical model with a collection of operations defined on that model. For instance, in our example of a list of numbers above, the model is a 'set' of pairs (n,p) where n is a number and p is a position in a list referring to first, second, ... p^{th} element positions. Using mathematical models such as sets means that their structures can be analysed and manipulated more easily and in a more rigorous form. For instance, the meanings and properties of set membership, union, intersection and other relevant operations on sets are very well defined and understood. Such properties facilitate the task of designing programs and verifying their correctness. Programming using abstract data types obviously adds a further level of abstraction to the design process.

Ideally, we would like to have a high-level language with all these abstract data types built in, in the same way that records and arrays are built into Pascal. The difficulty is, of course, that for most of these abstract data types there is no unique and efficient implementation. In different circumstances where the frequency of the operations are varied, different implementations may be used. Another problem is the variation in defining data types. For instance, a list of elements may be defined as a sequence with operations get_first and get_next to access all its elements *or* as a recursive data type with operations get_first and get_rest to access the first element and the rest of the elements as a list. Depending on the exact operations supported, size, form and other factors, there can be many implementations, each one suitable for a particular variation. Thus it is left to the programmer to define abstract data types and provide an implementation for them in terms of data structures, data types of a language (such as Pascal) and other abstract data types. In this way, a hierarchy of types can be defined so that each type is implemented in terms of the types lower in the hierarchy (see figure 2.2). In the remainder of this book we shall study how these implementations are carried out and how they depend on the specific contexts in which they are used. We will also illustrate how the implementation details of a data type can be decoupled from its abstract specification.

Example: Abstract Data Type Stack

Let us define and study an application of an abstract data type *stack*. Stacks are used in many places in computer science and are often known as Last In First

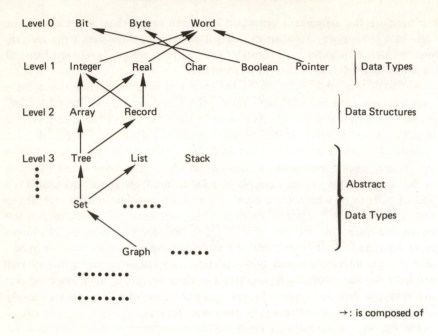

Figure 2.2   Hierarchy of data types.

Out (LIFO) lists. Roughly, a stack is a set of objects which arrive at different points in time and are added to the set. When elements are deleted from the set, the object which was inserted last is removed first.

A stack can be visualised as a pile of objects where objects can be added to or removed from the pile only from the top (see figure 2.3). In abstract terms, a stack should be of unlimited size. However, in real applications, we usually have a bound on the size of a stack. For this reason, in the following it is assumed that the size of a stack may reach this bound and therefore that stack becomes full.

An informal description of the ADT stack and its corresponding operations can be given as:

ADT stack: *set* of elements of type *elemtype*
(∗ elemtype is used to refer to the type of the individual elements in a stack. Elemtype can potentially be any defined type ∗)
Operations:

   procedure initialise(**var** S : stack);
           This procedure assigns an empty stack to S.

   procedure push(**var** S : stack; a : elemtype);
           Stack S should not be full. This procedure adds the element a at the top of the stack.

procedure pop(**var** S : stack);

> Stack S should not be empty. The top element of the stack is removed.

function top(S : stack) : elemtype;

> Stack S should be non-empty and this function returns the top element of the stack S. The stack S is left unchanged. (If elemtype is a structured type, this function should be rewritten as a procedure, see note 1 of section 3.1.)

function empty(S : stack) : boolean;

> This function returns true if S is empty and false otherwise.

function full(S : stack) : boolean;

> If S is full this function returns a true value.
> Otherwise it returns a false value.

The effects of these operations are shown diagrammatically in figure 2.4. Note that the type of the elements is specified as 'elemtype' which can be any type (character, record etc.). For any specific application, we need to specify exactly what elemtype is. However, note that the stack operations are independent of the type of the stack elements.

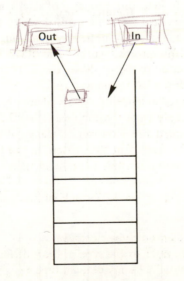

Figure 2.3 Pictorial model of a stack.

The type stack has now been defined (informally) in abstract form with no reference to any implementation. (The definitions are very intuitive. For rigorous definitions, see section 2.3.) Let us now look at how such an abstract data type can be used in some applications Perhaps the simplest application of a stack is

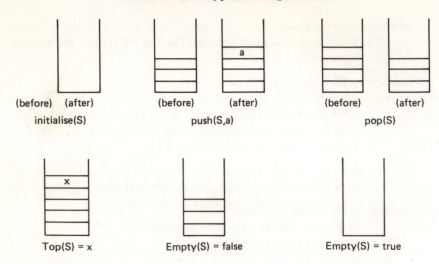

(before)   (after)   (before)   (after)   (before)   (after)

initialise(S)   push(S,a)   pop(S)

Top(S) = x   Empty(S) = false   Empty(S) = true

Figure 2.4   Pictorial representation of stack operations.

to reverse the elements of a sequence. We shall develop a program to read a sequence of characters delimited by a '.' character and print the sequence in reverse order. The 'Last In First Out' property of stacks makes them suitable for this problem. As the characters are being read, they are added to a stack. When the delimiter is reached the stack is emptied by 'removing' elements one by one. Figure 2.5 shows how this process works. The stack used for this problem

SEQUENCE: 'abcd'

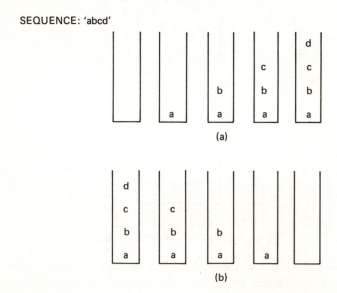

Figure 2.5   Operation of the 'reverse' program.

should be a stack of characters. Thus, the elemtype as used in the definition of stack should be 'char' A Pascal program which uses an ADT stack to reverse a sequence of characters is then listed.

'Reverse' Program

```
program reverse (input, output);
   var c : char;
       S : stack;              (* S is a stack of characters *)
   begin
     initialise(S);            (* clear the stack S *)
     read(c);                  (* read the first character *)
     (* while the delimiter is not reached and the stack is not
         full, read characters and push them into the stack S *)
     while (c <> '.') and not full(S) do
       begin
         push(S,c);                 (* add c onto the stack *)
         read(c)                    (* read the next character *)
       end;
     while not empty(S) do (* remove elements from the top of *)
       begin                  (* the stack until the stack is empty  *)
         write(top(S));
         pop(S)
       end
   end.
```

So far, the reverse program uses 'stack' as if it were a pre-defined type such as integers. The next stage in the development of this program is to implement the ADT stack. One way to implement a stack is to use an array to hold the elements (see figure 2.6).

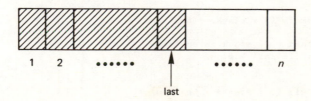

Figure 2.6 Array implementation of stacks.

Abstract Data Types and Algorithms

The variable 'last' indicates the top of the stack. Now, the type stack can be defined as

```
const n =    ;                  (* n is the maximum size of a stack *)
type  stack = record
                elements : array [1..n] of elemtype;
                last : 0..n    (* 0 signifies an empty stack *)
              end;
```

The elements of the stack are elements[1], elements[2], . . ., elements[last]. When the stack is empty, 'last' is equal to zero. Based on this data structure, the operations of a stack can now be defined.

```
procedure  initialise (var S: stack);
  var i : integer;
  begin
    for i := 1 to N do S.elements[i] := elem0;
    (* elem0 is an element of type elemtype *)
    S.last := 0  (* S is an empty stack *)
  end;

procedure  push (var S : stack; a : elemtype);
  begin
    S.last := S.last + 1;      (* increment top *)
    S.elements[last] := a      (* insert a into the stack *)
  end;

procedure  pop(var S : stack);
  begin
    S.last := S.last - 1       (* decrement last *)
  end;

function  top(var S : stack) : elemtype;
  begin
    (* return the top element of the stack *)
    top := S.elements[S.last]
  end;

function  empty (S : stack) : boolean;
  begin
    (* if last = 0, stack is empty *)
    empty := (S.last = 0)
  end;

function  full (S : stack) : boolean;
  begin
    (* if last = n, stack is full *)
    full := (S.last = n)
  end;
```

Note that the above implementation of stacks is general in the sense that the type of individual elements 'elemtype' can be changed to any type. For instance, for the 'reverse' program listed previously, 'elemtype' should be replaced with *char* (with the exception that if elemtype is a structured type, the function push should be written as a procedure — see note 1 of section 3.1). Also, note that in the procedure initialise, it is not strictly necessary to assign an arbitrary value to all the elements of the array. But, if this is not done, some Pascal compilers would reject 'write(empty(S))', since S contains undefined components.

Once an ADT is specified and implemented, it can be used like any other built-in type with no reference to its particular implementation. Before we can do so, however, it must be established that the implementation conforms to the specification. In other words, the implementation must be proved to be correct. It is not possible to give such proofs since our specifications are given informally in English. Section 2.3 discusses a formal language to express specifications and shows how proofs can be constructed.

Advantages of using Abstract Data Types

(1) Using ADTs implies abstracting the structure of the data and only highlight-
    ing the main characteristics. This process reduces the amount of detail that
    must be handled at any point in time. The combined effect of these features
    is the easier development of programs which are more susceptible to proofs
    and reasoning about their behaviours.
(2) Use of ADTs results in *modular* programs. As each ADT is specified and
    implemented, it can be defined as a *module*. Such modules can then be used
    in other modules or programs without being recompiled. Thus a lot of time
    and effort is saved when developing a large program. This mechanism also
    hides details of the implementation of each module, and only makes acces-
    sible the operations which can be performed on an ADT. Unfortunately, the
    Pascal language does not, in general, encompass the concept of a module.
    Other languages, such as Modula-2 and Ada, have facilities for defining
    modules. However, certain implementations of Pascal do provide some
    limited but useful mechanisms for defining modules. For instance, the SVS
    Pascal system provides the concept of a *unit* which is a set of declarations,
    procedures and functions and is the unit of compilation. Thus, for example,
    a unit for the ADT stack of characters can be defined as

```
unit  stk;
interface
(* all declarations in this section
   can be used outside this unit *)

const n =    ;  (* max. size of the stack *)
type stack = record
                elements : array[1..n] of char;
```

```
                    last : 0..n
                  end;
procedure initialise (var S : stack);
procedure push (var S : stack; a : char)
procedure pop (var S : stack);
function top (S : stack) : char;
function empty (S : stack) : boolean;
function full (S : stack) : boolean;

(* The end of the interface section *)

implementation
(* This section gives the implementation of all the
   procedures and functions specified in the interface
   above. In addition local variables, types, constants,
   labels, procedures and functions may be defined *)


procedure initialise;
  (* The parameters need not be specified again *)
  var i : integer;
  begin
    for i := 1 to N do S.elements[i] := '0';
    S.last := 0
  end;
  ...
  ...
  ...
function full;
  begin
    full := (S.last = n)
  end; (* of full *)
end.   (* end of the unit *)
```

The **interface** section defines global types, variables and other necessary declarations, procedures and functions which can be called outside this unit. The **implementation** section defines any local declarations and routines in addition to the implementation of procedures and functions defined in the interface section. Then, other units or programs can use this unit by means of a **uses** clause.

```
program reverse(input, output);
  uses  stk;    (* This makes available the interface
                    declarations of the unit stk *)
  var   c : char;
        S : stack;
  begin
    .....
  (*  as before  *)
    .....
  end.
```

The language Modula-2 provides somewhat similar mechanisms for defining and implementing data types. In Modula-2, a *definition module* gives the interface section and a separate *implementation module* gives the implementation of procedures and functions defined in the definition module. A major shortcoming of both these approaches is that the data structures which are used to implement an abstract data type must be given in the interface section (*definition module* of Modula-2). This, of course, contradicts the main objective of data abstraction – that is, hiding implementation issues from the application programs using abstract data types. In principle, such data structures need not be in the interface section, but translators for such languages need the extra information for efficient compilation. (Modula-2 allows a somewhat restricted form of data abstraction known as *opaque* types. But these may only be of type pointer.)

(3) Since the implementation of an ADT is hidden from the applications, different strategies can be used in the implementation without affecting the programs using that ADT. One particular advantage is that trade-offs between time and space can be made. Having designed a program using ADTs, the frequency of calling each operation can be determined and thus, when implementing the ADT, suitable data structures and/or other ADTs can be used to speed up the most frequently used operations. In examples later in the book we shall see exactly how these trade-offs can be made.

(4) Finally, since ADTs are mathematical objects, their meaning can be formally specified. Based on these specifications, it is then possible to verify that a given implementation of an ADT is correct and agrees with the specification. In the next section we shall establish the correctness of the array implementation of stacks. However, for the remainder of this book we shall rely on informal specification of ADTs and the proofs of their implementations.

2.3 Formal Specification of Abstract Data Types

Informally, we define a stack to be a set of objects. To be able to define the meaning of each operation we need to be precise about this set. Objects are inserted into a stack at different points in time and, when removing an object, the one which was inserted last is removed first. Thus the objects in this set are actually pairs of an object and a time-stamp corresponding to the object's arrival time. Thus, the stack of figure 2.4 could be the set {(a,1), (b,3), (c,5), (d,7)} where 1, 3, 5 and 7 are points in time when a, b, c and d were added to the stack respectively. So, the abstract data type *stack* can be defined as a set of pairs. For practical reasons a stack should have a maximum allowable length. To model the time-stamp, we can use integers to denote time. Since we do not necessarily need the real time, we can model points in time by an *increasing* sequence of integers. Therefore, for a stack, we also need to have an indication of the current time-point. It may be argued that such stacks are over-specified. A simpler specification for a stack would be an *ordered set* with no reference to time-stamps (see exercise **2.12**). However, to illustrate concepts such as invariances etc., we shall follow the former specification.

stack : : c	: integer	(* current time-point *)
max	: integer	(* maximum allowable size *)
elems	: pair-set	(* elems is a set of pairs *)
pair : : object	: elemtype	(* object is of type elemtype *)
t	: integer	(* t is a time-stamp *)

In the above, ':' is used to denote that a stack consists of c (current time point), max (maximum size) and elems which is a set of pairs. Each pair is composed of an object and a time-stamp. In the following we use the notation (a,b) for a pair with object = a and t = b.

For valid stacks we must ensure that no two objects in the pair-set can have identical time stamps. This is because not more than one object can be added to the stack at the same point in time. Therefore we need to specify a constraint on type stack.

$$\text{invariance 1: not } [\ \exists\ a,b,t\ (a,t) \in \text{elems and}$$
$$(b,t) \in \text{elems and}$$
$$a \neq b]$$

This means that this property is an *invariance* and must always be true about a stack. A second invariance must hold that c, the current time-stamp, be greater than all the time-stamps in the stack.

$$\text{invariance 2: } \forall\ (a,t) \in \text{elems}\ c > t$$

Finally, a last invariance should ensure that a stack may not have more than max objects.

$$\text{invariance 3: } |\text{elems}| \leqslant \text{max}$$

Now we specify each operation by giving its *pre* (condition) and *post* (condition) which must be true *before* and *after* the operation respectively. In post-conditions we use the 'prime' superscript ' to denote the value of a variable after the operation is executed. For instance, in procedure push, S and S' denote the stacks before and after the procedure has been carried out.

initialise(s)
| | pre : true | (* none *) |
| | post : elems' = { } | (* elems' is an empty set *) |

push(s,a)
| | pre : \|elems\| < max | (* s is not full *) |
| | post : c' > c and (a,c) ∈ elems' | (* the new time-stamp c' is larger than |

the previous time-stamp c′ and the
pair (a,c) belong to S *)

pop(s)

    pre   : | elems| > 0   (* s is not empty *)

    post : ∃ (a,b) ∈ elems such that   (* (a,b) is the element in s with largest
           [∀(p,q) ∈ elems b ⩾ q]   time-stamp b and is deleted from
        and elems′ = elems − {(a,b)}   elems *)

top(s)

    pre   : | elems| > 0

    post : ∃ (a,b) ∈ elems such that   (* (a,b) is the element in S with
           [∀(p,q) ∈ elems b ⩾ q]   largest time-stamp b and a is
        and top(s) = a   returned as the value of the
                                        function. Note that c and s are
                                        not changed *)

empty(s)

    pre   : true
    post : empty(s) = (| elems| = 0)

full(s)

    pre   : true
    post : full(s) = (| elems| = max)

To implement these operations, the abstract data type should be refined using other data types. We showed how a stack can be implemented using an array. Let us call this new data type a_stack (for array_stack).

    a_stack : : last   : 0 . . n   (* top of stack *)
               n   : integer   (* max size *)
             elements : array [1. .n] of elemtype (* the elements in
                                                   the stack *)

As for a stack, we can define each operation of the a_stack. For instance:

    push(s,a) :
      pre   : last < n
      post : last′ = last + 1 and
             elements′[last′] = a

To verify that this implementation is correct we must first define a mapping f which, given an a_stack, would give a corresponding stack. This can be easily expressed as

    f : a_stack → stack   (* f is a mapping defined from
                                  a_stack to stack *)

    f(x) = y

where x = (last, n, elements) and
y = (c, max, elems)
where c   = last + 1
max   = n
elems = {(elements[1], 1),
(elements[2], 2),

.

.

(elements[last], last)}

Using this mapping, various properties of this implementation and its correctness can be verified formally. As an example, let us consider the 'push' operation. We shall denote the 'push' operation for the stack and a_stack types as s.push and a.push respectively. To ensure that the implementation conforms to the specification, we need to show that performing a.push(s, a) has the same overall effect as performing s.push(f(s), a). We assume that each data type on its own is consistent — that is, the pre-condition and post-condition of each operation do not contradict the invariances of that data type. Informally, we need to show that, firstly, if pre(f(s), a) is true then pre(a.push(s, a)) is also true. Then it must be shown that if pre(s.push(f(s), a)) and post(a.push(s, a)) are both true, post(s.push(f(s), a)) must also be true. Formally, two things must be proved.

(a) $\forall s \in$ a_stack:

pre-condition(s.push(f(s), a)) $\Rightarrow$ pre-condition(a.push(s, a))

(b) $\forall s \in$ a_stack:

pre-condition(s.push(f(s), a) *and* post-condition(a.push(s, a))
$\Rightarrow$ post-condition(s.push(f(s), a))

To prove these statements:

Let s = (last, n, elements)   f(s) = X where
X = (last, n, elems) as above

(1) Now, pre-condition(s.push(f(s), a)) $\equiv$ | elems | < max

But | elems | = last
Under the mapping f, max = n
Thus | elems | < max $\Rightarrow$ last < n
Pre-condition(a.push(s, a)) $\equiv$ last < n
Therefore (a) is proved.

(2) pre-condition(s.push(f(s), a)) $\equiv$ | elems | < max
post-condition(a.push(s, a)) $\equiv$ (last' = last + 1 *and*
elements'[last'] = a)

$(|\,\text{elems}\,| < \max \; and \; \text{last}' = \text{last} + 1 \; and \; \text{elements}'\,[\text{last}'] = a)$
$\Rightarrow (\text{last}' < n \; and \; c' = c + 1 \; and \; (\text{elements}'\,[\text{last}'], \text{last}') \in \text{elems}')$
$\Rightarrow (c' > c \; and \; (a, c) \in \text{elems}')$
Post-condition $(s.\text{push}(f(s), a)) \equiv (c' > c \; and \; (a, c) \in \text{elems}')$
This proves (b) above.

(a) and (b) above thus ensure that the effects of the procedure push on an a_stack s is exactly the same as the effect of the procedure push on stack f(s).

The above proof should be applied to all the other operations of stacks. We have assumed that the a_stack implementation satisfies its own specifications. For this simple implementation it is trivial to verify the pre-conditions and post-conditions. However, for more complex and large implementations formal techniques should be used.

The above formal language to specify data types assumes the existence of a model for a data type and then specifies its invariances and the operations on that data type. It is believed that this approach is closer to implementation languages such as Pascal. However, there are formal specification languages which avoid assuming any model and specify the intended meaning of a data type and its operations via algebraic equations. This approach is briefly discussed in the Bibliographic notes at the end of this chapter.

As we mentioned before, we shall be content with informal specifications and proofs in this book. The emphasis will be on definition of data types and how efficient implementations can be developed and used in various application areas. A substantial program using a few abstract data types is given in the case study of section 3.4 which involves reading, storing and evaluating arithmetic expressions.

Exercises

2.1. What is 'abstraction'? Give a few examples of abstraction in every-day life.
2.2. Differentiate between data types, data structures and abstract data types.
2.3. Recursive procedures and functions calls are sufficiently powerful control structures to express any loop structures of Pascal, and vice versa. Give the equivalent forms of each loop construct of Pascal using only procedures and functions, and vice versa.
2.4. Data abstraction and operation abstraction can be achieved (to some extent) in a high-level language. How can control abstraction be accommodated in a high-level language such as Pascal? For example

For all x **in** X do S(x);
(∗ X is a set ∗)
(∗ Meaning: For every x in the set X, execute the procedure S ∗)

2.5. Define pre-conditions and post-conditions of the search and b_search routines of chapter 1. Are the two routines equivalent?

2.6. Show that the invariance of the ADT stack is not violated by its operations; that is, the three invariances are satisfied before and after each operation.

2.7. Show that

$$\forall\ x \in a\_stack\ \exists\ s \in stack\ such\ that\ s = f(x)$$

That is, f is defined for all a_stacks.

2.8. Show that

$$\forall\ s \in stack\ \exists\ x \in a\_stack\ such\ that\ f(x) = s$$

That is, for each valid stack, there is *at least one* a_stack.

2.9. Specify the pre-conditions and post-conditions of the a_stack operations.

2.10. Using the specifications of exercise **2.9**, show that a_stack operations satisfy stack specifications.

2.11. Show that

```
begin
    push(s, a);
    A := top(s);
end;
```

has post-condition $\equiv (A = a)$.

2.12. Let E denote Elemtype. Then powers of E will be defined as

$$E^0 = \{\ \}$$
$$E^1 = E$$
$$E^2 = E * E$$
$$E^3 = E * E * E$$
etc.

The iteration over E can now be defined as

$$E^* = E^0\ U\ E^1\ U\ E^2\ U\ E^3\ U \ldots U\ E^n$$

Hence we can define a (bounded) stack as

$$stack :: \quad n \quad : integer$$
$$elems : E^*$$

Using this model for a stack, specify the pre-conditions and post-conditions of stack operations and therefore show that the array implementation of stacks conforms to that specification.

Bibliographic Notes and Further Reading

The ideas of stepwise refinement and structured programming which primarily make use of procedural abstraction can be found in Wirth (1971) and Dahl *et al.* (1972).

The concept of ADTs is not new. It originated from the class concept of the language Simula (Birtwistle *et al.*, 1973). Recently it has gained popularity for software development and system design (Liskov *et al.*, 1977; Shaw *et al.*, 1977; Shipman, 1981). The formal specifications of ADTs can also be expressed algebraically (Guttag and Horning, 1978) with no reference to the domain of the ADTs. The book by Martin (1986) is a good textbook on algebraic specification for most of the ADTs discussed in this book. The technique used in this chapter is model oriented. This approach first defines a data type explicitly as an abstract model (for example, set, sequence etc.) and then the operations in terms of this model. These techniques are further studied in Jones (1980). In the former approach, a data type is defined implicitly by relating its operations via axioms. For instance, the ADT stack can be defined algebraically as

Syntax : empty : stack $\rightarrow$ boolean
initialise : $\rightarrow$ stack
push : stack * elemtype $\rightarrow$ stack
pop : stack $\rightarrow$ stack U {error}
top : stack $\rightarrow$ elemtype U {error}

Axioms : empty(initialise) = true
empty(push(s, a)) = false
top(push(s, a)) = a
top(initialise) = error
pop(push(s, a)) = s
pop(initialise) = error

For an empty stack, operations pop and top return an error. For more details see Guttag *et al.* (1978) and Martin (1986). Many programming languages have facilities for defining ADTs, among which are Modula-2 (Wirth, 1982) and Ada (Wiener and Sincovec, 1983). The book by Ford and Wiener (1985) summarises the differences between Pascal, UCSD Pascal and Modula-2 in this respect.

Usually when an ADT is defined, *multiple instances* may be cited. For instance, many instances of the stack data type may be defined in a similar way to integer variables:

s1,s2 : stack;
i,j,k : integer;

In certain circumstances where *only one instance* of an ADT is required, it may be advantageous to define that instance as a global variable of the ADT, and its operations need not have an explicit parameter to refer just to this instance. For example, if only one stack is required in an application, we can define its operations as

procedure initialise;
procedure push(a : elemtype);
procedure pop;

  function top : elemtype;
  function empty : boolean;

In chapter 7, the ADT graph is defined for just one instance.

A final note is in order about *generic data types* and *polymorphic data types*. At present, if we need a stack of integers and a stack of characters, we have to duplicate the Pascal code and the data declarations for the two stacks, and also to give different names to the operations on the two stacks. This is so even though it is obvious that stack operations are unaffected by the type of its elements. Polymorphic or generic data types refer to objects and operations having more than one type. For instance, if 'stack(a)' refers to the type 'stack of elements of type a', then the type of operation 'top' could be specified as 'for all a (stack(a) → a)' which signifies that type of pop is a function which takes a stack of 'a' and returns a value of type 'a'. Furthermore, this is so for any type 'a'. ML (Milner, 1978) and HOPE (Burstall *et al.*, 1980) are among the few languages that provide polymorphic types.

Birtwistle, G. M. *et al.* (1973). *SIMULA Begin*, Auerback Press, Philadelphia, Pennsylvania.

Burstall, R. M., MacQueen, D. B. and Sannella, D. T. (1980). 'Hope: an experimental applicative language', in *Proceedings of 1980 Lisp Conference, Stanford, California*, pp. 136–143.

Dahl, O. J., Dijkstra, E. W. and Hoare, C. A. R. (1972). *Structured Programming*, Academic Press, London and New York.

Ford, G. A. and Wiener, R. S. (1985). *Modula-2: A Software Development Approach*, Wiley, Chichester and New York.

Guttag, J. V. and Horning, J. J. (1978). 'The algebraic specification of abstract data type', *Acta Informatica*, Vol. 10, pp. 27–52.

Guttag, J. V. *et al.* (1978). 'Abstract Data Types and Software Validation', *CACM*, Vol. 21, No. 12, December, pp. 1048–1064.

Jones, C. B. (1980). *Software Development: A Rigorous Approach*, Prentice-Hall, Englewood Cliffs, New Jersey.

Liskov, B. *et al.* (1977). 'Abstraction mechanisms in CLU', *CACM*, Vol. 20, No. 8, pp. 564–576.

Martin, J. J. (1986). *Data Types and Data Structures*, Prentice-Hall, Englewood Cliffs, New Jersey.

Milner, R. (1978). 'A theory of type polymorphism in programming', *Journal of Computer and Systems Sciences*, Vol. 17, No. 3, pp. 348–375.

Shaw, M. *et al.* (1977). 'Abstraction and verification in ALPHARD: defining and specifying iteration and generators', *CACM*, Vol. 20, No. 8, pp. 553–563.

Shipman, D. W. (1981). 'The functional data model and the data language DAPLEX', *ACM TODS*, Vol. 6, No. 1, March.

Wiener, R. and Sincovec, R. (1983). *Programming in ADA*, Wiley, New York.

Wirth, N. (1971). 'Program Development by Stepwise Refinement', *CACM*, Vol. 14, No. 4, pp. 221-227.
Wirth, N. (1982). *Programming in Modula-2*, Springer-Verlag, New York.

3 Elementary (Linear) ADTs

In chapter 2 we defined an abstract data type as a model with a set of operations. We can broadly classify ADTs according to the structure of their models. The first class of ADTs includes those whose models have a *linear structure*. A linear structure is one in which each element (except the last) may have exactly one *successor* element and each element (except the first) one *predecessor* element. The ADT stack falls into this category. The model for a stack is a *set of elements* where the elements are time-stamped on their arrival into the set. According to the time-stamps, the elements can be organised into a linear structure. This is why there can be no more than one pair with similar time-stamps in a stack.

Trees and graphs are examples of non-linear data types. In trees, each element is allowed to have a maximum of one 'previous' element but many 'next' elements. In graphs, an element can have many 'next' and 'previous' elements. In this chapter we shall study three linear ADTs, namely stacks, queues and lists.

Figure 3.1 illustrates linear and non-linear structures.

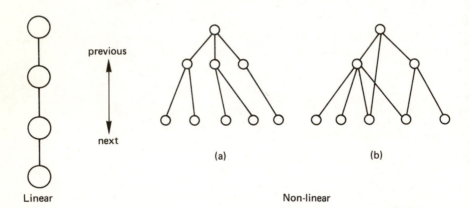

Figure 3.1 Linear and non-linear structures.

3.1 The ADT Stack

The structure and operations of stacks were described in detail in chapter 2. An implementation of bounded stacks was given where the maximum number of elements in the stack was known in advance. For unbounded stacks, an implementation using pointers and dynamic storage is preferred.

48

Pointer Implementation of Stacks

Using pointers a stack can be implemented using a linked list of cells. Figure 3.2 shows the pointer representation of a stack S with elements a, b, c and d pushed into the stack in that order. Thus, we can declare the following types:

```
type  cellptr = ↑cell;
      cell = record
               element : elemtype;
               next : cellptr
             end;
      stack = cellptr;
```

The elements of the stack are given the generic type 'elemtype' which can be replaced with any data type as required.

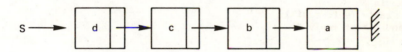

Figure 3.2   Pointer representation of a stack S.

An empty stack S is thus represented by a *nil* pointer. Using the above declaration the operations on stacks can be defined accordingly:

```
procedure initialise (var s : stack);
  begin
    s := nil
  end; (* of initialise *)

procedure  push (var s : stack; a : elemtype)
  var p : cellptr;
  begin
    new (p);                      (* create a new cell for a *)
    p↑.element := a;
    p↑.next  := s;
    s := p
  end; (* of push *)
```

```
procedure pop (var s : stack);
  var p : cellptr;
  begin
    if  s = nil then error        (* if stack s is empty *)
    else
        begin
          p := s;
          s := s↑.next;
          dispose(p)
        end
  end; (* of pop *)

function top (s : stack) : elemtype;
  begin
    if  s = nil then error
    else
        top := s↑.element
  end; (* of top *)

function empty (s : stack) : boolean;
  begin
    empty := (s = nil)
  end; (* of empty *)
```

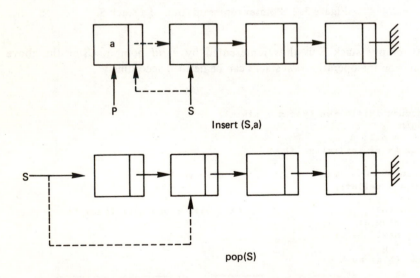

Insert (S,a)

pop(S)

Figure 3.3 Procedures Insert and Pop on stacks.

In figure 3.3 the dotted lines show the pointers which need to be modified.

Notes

1. In Pascal a function cannot return a value other than a simple-type or a pointer. Thus, if elemtype is a large record or an array structure, either a pointer to the structure should be returned or the function should be re-written as a procedure. For instance, the function top on stacks can be declared as

    procedure top (s : stack; **var** a : elemtype)

2. The function 'full' may not be implementable in standard Pascal. When calling the *new* procedure, some Pascal systems would return a status which indicates whether or not any more storage is available. Most systems would return a nil pointer in this situation.

3. The procedure *error* reports an error message to the user and exits the program.

4. The assignment operator ':=' can be used with abstract data types, but it must be used with care. This is particularly true when Pascal pointers are used. For instance, let us consider the following example:

    var s1,s2 : stack;
    ......
    s1 := s2;

The assignment statement should copy s2 into s1 with no other side-effects. However, since the type stack is a pointer, this assignment statement makes the pointer s1 point to the beginning of the stack s2. Thus if, later on, stack s1 is modified, this will be reflected in s2 as well! To overcome this, we can define an assignment operation 'assign' on stacks, which is then implemented properly by copying the entire structure of s2 into s1:

```
procedure assign(var s1 : stack; s2 : stack);
  begin
    if s2 = nil then s1 := nil
    else begin
         new(s1);
         s1↑.element := s2↑.element;
         assign(s↑.next, s2↑.next)
       end
  end; (* of assign *)
```

This anomaly of pointer implementations will hold for all the ADTs to be discussed in this book and, therefore, we shall always assume a specialised 'assign' operation for each ADT. (Similar problems occur when pointers are represented by indices in an array.)

In the rest of this section we shall look at two more examples of how stacks are used.

Reverse Polish Notation

An interesting application of stacks involves reading and transforming arithmetic expressions into a normal form known as Reverse Polish (RP). Later on, we shall look at the problem of evaluating such expressions. This is a kind of operation which a compiler needs to do when an arithmetic expression is encountered.

The usual notation used for writing expressions is 'in-fix' (or 'in-order'); that is, the operators come in the middle of two operands such as $(a + b)$ etc. To remove ambiguities in complex expressions, brackets are used to group sub-expressions. Operators are given precedence so that '*' and '/' have higher precedence than '−' and '+'. Thus

$$a + b * c \equiv a + (b * c)$$
$$a * b \: / \: c \equiv (a * b) \: / \: c$$

The first step towards evaluating such expressions is to transform them into a form without brackets. One such notation is known as 'post-fix' or 'post-order' and is usually referred to as 'Reverse Polish' after the Polish mathematician who invented it. In Reverse Polish (RP), the two operands come first, and then an operator follows. For instance, $(a + b)$ is represented as '$a \: b \: +$'. The above two expressions are thus:

$$a \: b \: c * +$$
$$a \: b * c \: /$$

Note that

$$(a + b) * c \quad \equiv a \: b + c *$$
$$(a * (b \: / \: c)) \equiv a \: b \: c \: / \: *$$

Now, we are ready to develop a program to perform this transformation from in-fix to post-fix form. In the following, let us assume that the operands are represented by lower-case letters and each in-fix expression is terminated by a ';'. We may assume the following precedence of operators:

$$* \: /$$
$$+ \: -$$
$$)$$
$$;$$
$$@$$

lower precedence

The symbol '@' is used to indicate the beginning of an expression (see later). Let us consider the following example:

$$(a + b * c) * (p - q);$$

When transforming expressions into Reverse Polish, it should be noted that operands come first and then an operator appears. Thus, when reading the in-fix expressions from left to right the operands can be output immediately, but the operators should wait until their corresponding operands are output. As a result, the operators should be stored in a data structure until they can be output in their proper place. Furthermore, when a new operator is processed, if it is of higher precedence than the previous one, then the latter should wait until this new operator is output. Such a constraint enforces a LIFO structure on the data structure and therefore makes it a stack. An outline of this process may be summarised as:

```
initialise a stack S;
repeat
    read the next symbol (symbol)
    case symbol of
    operand  : output symbol
    operator :
      while (* this operator is of lower or equal precedence
              to its previous operator (i.e. top(S)) *) do
          begin
             output top(S);
             remove top(S);           (* i.e. pop(S) *)
             (* now add the operator symbol to
                 the stack for later processing *)
             push(S,symbol)
          end;
    open bracket  :
      (* push it onto the stack to mark the
          beginning of a subexpression *)
    close bracket :
      (* output all the operators on top of the stack
          until a matching open bracket is encountered *)
until end of expression
```

To ensure correct operation of this algorithm, we initialise the stack with a '@' symbol and when a ';' is found all the operators are output until the '@' is encountered. The following is a Pascal version of the algorithm. The 'elemtype' of stack should be specified as 'char'.

```
S : stack;          (* of char *)
token : char;

initialise(S);
push(S, '@');       (* To mark the beginning of the stack *)
repeat
      read(token);
      if token in ['a'..'z'] then write(token)
      else
```

```
     case token of
       '(' : push(S, token);
       ')' : begin
               while top(S) <> '(' do (* to match the      *)
                   begin                (* corresponding '(' *)
                       write(top(S));
                       pop(S)
                   end;
               pop(S)    (* to remove '(' *)
             end;
  ';','*',
 '+','-','/' : begin
                 while less_or_equal_prec(token, top(S)) do
                     begin
                         write(top(S));
                         pop(s)
                     end;
                 push(S, token)  (* push the operator *)
               end              (* into the stack    *)
       end; (* for case *)
until token = ';';
```

The less_or_equal_prec(a,b) is a function which returns true if a has lower or equal precedence to b (see section 3.5).

To show the mechanics of this algorithm, figure 3.4 lists the content of the stack S and the action and output at each stage.

EXPRESSION $\equiv$ (a + b + c) * (p − q);

Stack	Next_symbol	Action	Output
@	(	push	
@(	d	write	a
@(	+	push	
@(+	b	write	ab
@(+	*	push	
@(+*	c	write	abc
@(+*	)	pop	abc*+
@	*	push	
@*	(	push	
@*(	p	write	abc*+p
@*(	−	push	
@*(−	q	write	abc*+pq
@*(−	)	pop	abc*+pq−
@*	;	pop	abc*+pq−*
@;			

Figure 3.4 The working of the Reverse Polish algorithm.

Evaluation of Reverse Polish Expressions

Given a Reverse Polish expression, the next task is to evaluate it. RP expressions are particularly attractive since there are no brackets. According to the definition of RP, operators come after their operands. To evaluate an expression, we scan the expression from left to right. If the symbol is an operand we push it into a stack, and if the symbol is an operator we apply it to its operands which by then must be on top of the stack. The result must be pushed into the stack. At the end, the only operand left in the stack will be the result of the expression.

```
initialise(S);              (* S is a stack of integer *)
while more symbol do
    begin
        read a symbol
        case symbol of
        operand  : push(S, symbol);
        operator : begin
                    (* get the two operands *)
                    operand1 := Top(S); pop(S);
                    operand2 := Top(S); pop(S);

                    (* apply the operator to operand1 and operand2
                       and push the result into the stack, i.e. *)

                    case type of symbol of
                        plus   : result := operand1 + operand2;
                        mult   : result := operand1 * operand2;
                        minus  : result := operand2 - operand1;
                        divide : result := operand2 div operand1
                    end; (* of case *)
                    push(S, result);
                end
        end (* of case *)
    end; (* of while more symbol *)
writeln(top(S));
```

Notes

1. In the above program it is assumed that the letters representing operands are substituted by their actual values (see the case study later in this chapter).
2. This algorithm is best suited for computers with only one register. For multi-register machines a tree-representation of the expressions would be preferred (see chapter 4).
3. The above algorithm *interprets* the expressions. The algorithm can be modified to compile an expression into machine code (see exercise 3.7).
4. The above two algorithms do not deal with any error conditions.

Recursive Procedure Calls of High-level Languages

In high-level languages such as Pascal, procedures (and functions) may well call themselves. Such procedures are recursive. To ensure correct operation of recursive procedures, each time a procedure calls itself the local variables of that procedure is saved, together with a return address. Hence, when the called procedure is terminated, the initial procedure can be continued from the point after the recursive call. Because of these nested cells, the local information saved at each activation of a procedure should be stored in a stack. Here is an example to clarify these points.

```
procedure factorial (n : integer; var result: integer);
  begin
    if n < 0 then error
    else if n <= 1 then result := 1
        else  begin
                factorial(n-1, result);
                result := result * n
              end
  end; (* of factorial *)
```

The stack used to control the recursive calls when factorial(3, result) is called, would grow and shrink as

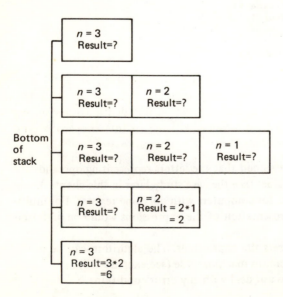

Recursive procedures usually result in a very elegant and concise description of a task. To offset this, the cost of manipulating the stack is incurred. For most problems, however, the simplicity of recursion is far outweighed by the loss of efficiency. In cases where efficiency is crucial, a recursive procedure can be written without recursion using an explicit stack. For the factorial procedure, an iterative solution can easily be defined without using a stack at all. However, to illustrate the techniques we shall give the non-recursive equivalent of the factorial procedure. The local information of each procedure call are variables n and result. Thus, we shall use a stack of integers. The recursion is to be replaced with a loop. A call to the procedure is changed to pushing the local information plus the return address (after the call). At the end of the procedure, we then take the top of the stack which tells us where to go (return address) and the new local information.

```
procedure factorial (n : integer; var result : integer);
   label 100;
   var s : stack;      (* of integer *)
   begin
      initialise(s);
100 : if n < 0 then error
      else if n <= 1 then result := 1
            else
               begin
               (* to save the local variables *)
                  push(s, n);
                  n := n-1;
                  goto 100
               end;
      while not empty(stack) do
         begin
            result := result * top(s);
            pop(s)
         end
   end; (* of factorial *)
```

In the above example there is only one recursive call, and therefore there is no need to store the return address, since there is only one such address (the beginning of the **while** loop). Also, the algorithm is simplified slightly by not storing 'result' in the stack but using it as a global variable (within the procedure). (Exercise **3.5** shows a more complex example of elimination of recursion).

A special case of a recursive procedure is when the recursive call is the last statement of the procedure. Such a procedure may look like

```
procedure p(......);
   begin
      ..
      ..
      ..
      p(......)
   end; (* of p *)
```

This type of recursion is known as 'tail' recursion. After the recursive call to p, the values of the local variables are not changed, the recursive call can be changed to a jump (**goto**) to the beginning of the procedure. However, care should be taken when changing the values of the variables corresponding to the parameters of the recursive call.

```
   procedure p(......);
      label 100;
      begin
100:    ..
        ..
        ..
      (* update variables corresponding
         to the value parameters of p *)
      goto 100
      end; (* of p *)
```

3.2 The ADT Queue

The abstract data type queue is a linear data type and its behaviour can be informally described as a First In First Out (FIFO) list. That is, the object which was entered in the queue first is retrieved first. Like stacks, the model for a queue is a set of objects which are time-stamped according to their arrival times. A pictorial representation of queues is given below, where objects enter at one end and are taken at the other end.

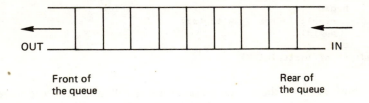

In a similar way to stacks, we can define the operations on a queue as

procedure initialise (var Q : queue);
  (* The queue Q is initialised to an empty queue *)

procedure enqueue (var Q : queue; a : elemtype);
  (* The element a is added to the Q at the rear of it. *)

procedure dequeue (var Q : queue);
  (* The element at the front of the queue is removed. Q should
    not be empty *)

function empty (var a : queue) : boolean;
  (∗ Empty is set to true if Q is empty and to false otherwise. ∗)

function front (Q : queue) : elemtype;
  (∗ Q should not be empty. The element at the front of
    the queue is returned. ∗)

Queues are primarily used in simulation applications where the behaviour of a system is simulated. For instance, simulating the working of a bank teller, customers form queues and they are served on a first-come, first-served basis. In other simulation applications, events which should be processed are queued and later on are processed one at a time. For instance, simulating the behaviour of traffic at a junction needs a queue of events such as 'a car turning right' etc. (see exercise **3.17**). In computer systems, where input/output devices are shared among many users, the requests are usually queued so that they can be processed later.

Implementation of Queues using Pointers

Since elements are inserted and deleted at two ends of the queue, we need to maintain the positions of the front and the rear of a queue. For this purpose we can allocate two pointers, *rear* and *front*, to the appropriate ends of a queue. Figure 3.5 shows the pointer representation of an empty queue and a queue of elements *a*, *b*, *c* and *d* arrived in that order.

Note that a dummy cell is used at the front of the queue. Such a dummy cell facilitates the manipulation of the linked list when the list is empty. Using this structure, we can now give the necessary declarations and the operations:

```
type cellptr = ↑cell
     cell    = record
                    element : elemtype;
                    next : cellptr
               end;
     queue = record
                front, rear : cellptr
             end;

procedure initialise (var Q : queue);
  begin
    new (Q.front);          (∗ create the dummy cell ∗)
    Q.rear := Q.front
    Q.rear↑.next := nil
  end; (∗ of initialise ∗)

procedure enqueue (var Q : queue; a : elemtype);
  var p : cellpt;
  begin
    new(p);
```

```
      p↑.next := nil;
      p↑.element := a;
      Q.rear↑.next := p
      Q.rear := p
   end; (* of enqueue *)
```

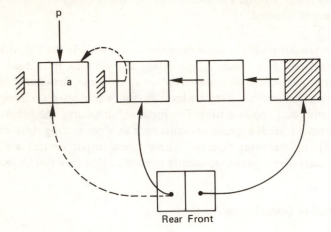

```
procedure dequeue (var Q : queue);
   var p : cellptr;
   begin
      if not empty(Q) then
         begin
           p := Q.front↑.next;
           if Q.rear = p then        (* when Q has only one element *)
              begin
              Q.front↑.next := nil;
              Q.rear := Q.front
              end
           else Q.front↑.next := p↑.next;
           dispose(p)
           end;
      end; (* of dequeue *)
```

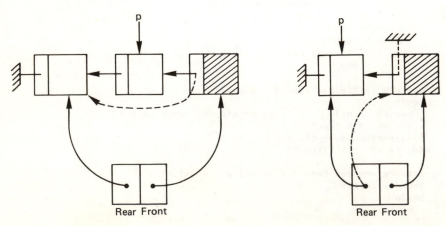

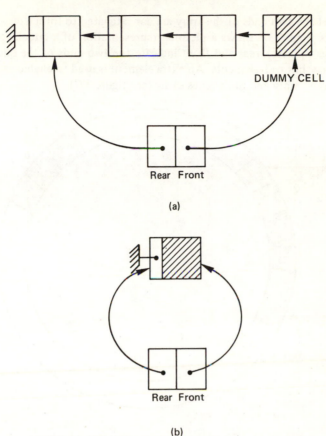

Figure 3.5 Pointer representation of queues: (a) non-empty queue; (b) empty queue.

Note that in dequeue, in the case when there is only one element in the queue, special action must be taken. The reader should verify that the operations enqueue and dequeue work for an empty queue. The operations front and empty are similar to those of a stack.

Array Implementation of Queues

When there is a bound on the maximum size of a queue, an interesting implementation using arrays can be used. If the maximum size of the queue is n, the elements can be stored in an array. However, because of dequeue operations we must ensure that the deleted elements can be used again. To achieve this, we can imagine that the two ends of the array are concatenated to form a circular array of elements. Figure 3.6 shows a pictorial representation of a circular array and two pointers (indices), *rear* and *front* indicate the two ends of the queue. Note that the array has $n+1$ elements. An extra element is used to enable us to distinguish between empty and full queues easily (see figure 3.7).

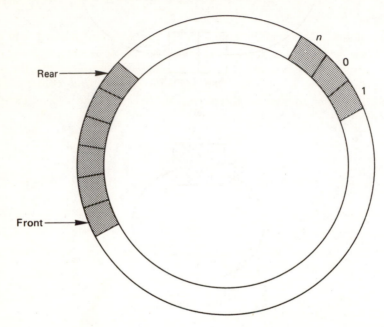

Figure 3.6   Circular array representation of queues.

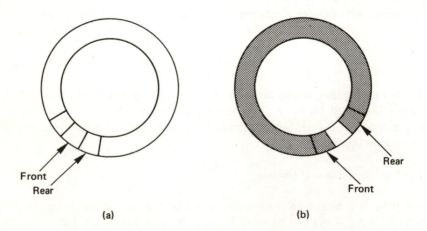

Figure 3.7   Empty (a) and full (b) queues.

Now we can give the following type declaration for the **ADT** queue:

```
type queue = record
                elements : array[0..n] of elemtype;
                front, rear : 0..n
            end;
```

The two pointers, *front* and *rear*, are simply two numbers indexing into the array. These two pointers advance in a cyclic fashion. Inserting a new element involves advancing the rear pointer and assigning the new element to the new location. Deleting an element involves only advancing the front pointer. First, we write a function to establish this circular movement of the pointers (indices of the array).

```
function increment (i : integer) : integer;
  begin
    increment : = (i+1) mod (n+1)
  end;
```

The function increment(i) returns the next location in the array. Now, the operations can be written very simply:

```
procedure initialise(var Q : queue);
  var i : integer;
  begin
    Q.front := 1;   (* An empty queue *)
    Q.rear := 0;
    for i := 0 to n do Q.elements[i] := elem0
    (* elem0 is an element of type elemtype *)
  end;

procedure enqueue(var Q : queue; a : elemtype);
  begin
    if not full(Q) then   (* Q should not be full *)
      begin
        Q.rear := increment(Q.rear);
        Q.elements [Q.rear] := a
      end
    else error
  end; (* of enqueue *)

procedure dequeue (var Q : queue);
  begin
    if not empty(Q) then
      Q.front := increment(Q.front)
    else error
  end; (* of dequeue *)
```

```
function front(Q : queue) : elemtype;
  begin
    if not empty(Q) then
       front : = Q.elements[Q.front]
    else error
  end; (* of front *)

function empty(Q : queue) : boolean;
  begin
     empty := (Q.front = increment(Q.rear)
  end; (* of empty *)

function full(Q: queue) : boolean;
  begin
    full := (increment(increment(Q.rear)) = Q.front)
  end; (* of full *)
```

Note that the function full with the usual meaning is included. Procedure error simply reports an error to the user and exits the program.

3.3 The ADT List

The stack and queue are two linear structures where insertions and deletions occur only at one or the other end of the structure. The abstract data type *list* is a generalised linear structure without these restrictions.

The ADT list is a sequence of elements $a_1, a_2, \ldots, a_{n-1}, a_n$. Each element has a particular position in the list. Positions are ordered in sequence from the first position to the last position in the list. The operations on a list should allow one: to insert and delete elements at any position in the list; to retrieve elements at any position (similar to top(stack) and front(queue)); to locate a given element in the list; and to move back and forth along the list. In fact, there are many ways in which the ADT list can be defined. To decouple the ADT list fully from its implementation we shall introduce an abstract notion of position rather than using numbers $1, 2, \ldots, n$ as used in the mathematical definition of a list. To simplify the operation of insert for a list with n elements, we define $n+1$ positions where the last position ($n+1$) refers to the end of the list. In this way, new elements can be added anywhere in the list. (The alternative would be to define two operations insert_before and insert_after a given position.) The set of operations for the ADT list can now therefore be given:

procedure initialise (var L : list);
   (* This sets L to be an empty list *)

procedure insert (var L : list; a : elemtype; p : position);
   (* The list $a_1, a_2, \ldots, a_p, \ldots a_n$ is transformed
      to $a_1, a_2, \ldots, a_{p-1}, a, a_p, a_{p+1}, \ldots, a_n$ *)

procedure delete (var L : list; p : position);

(* The list $a_1, \ldots, a_{p-1}, a, a_p, a_{p+1}, \ldots, a_n$

is transformed to $a_1, a_2, \ldots, a_{p-1}, a_{p+1}, \ldots, a_n$.

Note that after an insert or delete operation is executed, the values of any position variable in that list may no longer be valid. If necessary, we could specify that after insert(L,a,p), p will refer to element a and after delete(L,p), p will refer to position next(p). However, this requires that p should be defined as a var parameter. *)

function retrieve (p : position; L : list) : elemtype;

(* This function returns the element at position p of list L. *)

function next (p : position; L : list) : position;

(* This function returns the next position after p in list L. If p is the last position in the list, an error should be reported. *)

function previous (p : position; L : list) : position;

(* This function returns the previous position to p in list L. If p is the first position, an error is generated. *)

function first (L : list) : position;
function last (L : list) : position;

(* These two functions return the first and last positions in the list L respectively. Last refers to the position after the last element of the list. *)

function empty (L : list) : boolean;
function full (L : list) : boolean;

(* These two functions return a true value if the list L is empty and full respectively. *)

It should be noted that any other operation required can be written using these operations. As an example, we shall give the operation *search* which would find the first occurrence of a given element.

```
function search (a : elemtype; L : list) : position;
  (* find the first position p such that retrieve(p,L) = a *)
  var p : position;
  begin
    p := first(L);
    while (p <> last(L)) and (retrieve(p,L) <> a) do
        p := next(p,L);
    (* if a is not found, last(L) is returned *)
    search := p
  end; (* of search *)
```

Many applications using lists do not need to use all the operations listed above. So, for each particular application, different implementations can be given so that the operations which are needed most frequently are made more efficient. We shall discuss different implementation strategies and comment on their suitability for the list operations.

Pointer Implementation of Lists

Using cells similar to those used for stacks and queues, we can represent a list L
using pointers as follows:

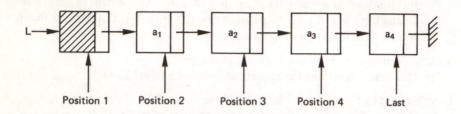

Position 1   Position 2   Position 3   Position 4   Last

Figure 3.8   Pointer representation of a list L.

The abstract type position is represented as a pointer to the previous cell to the
cell at that position. This is necessary since when inserting at position p, we need
access to the previous cell to p. This shall be apparent when we discuss the *insert*
operation. A dummy cell is used at the beginning to facilitate manipulating
empty lists.

```
type list = cellptr;
     position = cellptr;
 ........
procedure insert (var L : lists a : elemtype; p : position);
  var q : cellptr;
  begin
    new(q);
    q↑.element := a;
    q↑.next := p↑.next;
    p↑.next := q
  end; (* of insert *)
```

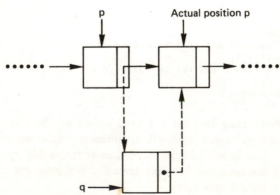

The reader should verify that this procedure works for cases where the list is empty (in this case p can only be last(L)), p = first(L) or p = last(L).

```
function first(L : list) : boolean;
  begin
    if not empty(L) then   (* if L is empty, position 1 does not exist *)
      first := L          (* only last(L) would be defined *)
    else error
  end;

function retrieve ( p : position; L : list) : elemtype;
  begin
    retrieve : = p↑.next.element
  end;
```

Operations initialise, delete, next, first, empty and full can be written efficiently using this structure. (The function full depends on the particular system of Pascal being used.) The functions previous and last can only operate by scanning the list from the beginning. For instance, previous can be written as:

```
function previous (p : position; L : list) : position;
  var q : position;
  begin
    previous := L;
    if p = first(L) then error
    else begin
           q := L;
           while q↑.next <> p do
               q := q↑.next;
           previous := q
         end
  end; (* of previous *)
```

This function has the maximum time requirement $O(n)$ where n is the size of the list. Similarly, last(L) needs $O(n)$ time.

If these two functions are used frequently in an application, a different implementation known as double-linked lists can be used. A list L would then be represented as

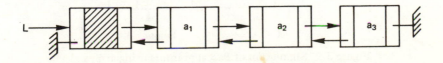

The function last can be implemented by maintaining a pointer to the end of the list. A similar method is by using a circular double-linked list (see exercise **3.8**). A possible type declaration for a doubly-linked list is

```
type cellptr = ↑cell;
     cell = record
                element : elemtype;
                next, previous : cellptr
            end;
     list = cellptr;
```

Array Implementation of Lists

A simple array representation of a list is very restrictive in terms of the operations which can be performed efficiently. The elements of a list L can be stored in an array as

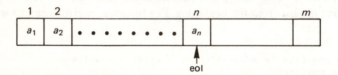

The eol index (an integer variable) indicates the end of the list. Obviously, m is the maximum size of the list. In this implementation the operations insert and delete are only efficient if they are applied to the beginning or the end of the list. If, for instance, an element is to be inserted in the middle of the list, all the elements to the right of that position must be shifted by one position to the right to make room for the new element. Such an operation needs a maximum time of $O(n)$. For the search function, an ordered array needs a time of $O(\log n)$ and an unordered array $O(n)$ (see chapter 1). All other operations are, however, relatively easy and efficient.

To alleviate some of the above problems, a second array implementation may be used. This implementation, in fact, is a simulation of the pointer implementation. Therefore, its use is only sensible if the language being used does not support pointers. Figure 3.9 shows a possible representation of the list $L \equiv a_1, a_2, a_3$.

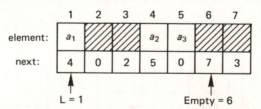

Figure 3.9 Simulation of Pascal pointers using arrays.

Each array location consists of an element and a next pointer. The list L is an index into the array to the first element of the list (a_1). Then the next field of this element gives the index of the next element (4). That is, a_2 is stored in the 4th location of the array. A zero indicates the end of the list. To be able to insert new elements, a mechanism is needed to find an empty location in the array. This can be done by maintaining another list called EMPTY which contains all the empty locations (in this example, locations 6, 7, 3 and 2). Since this structure closely mimics the behaviour of pointers, the operations are very similar to those of the pointer implementation of Pascal.

Note that, as for pointer implementations, a dummy location of the array can be used to denote an empty list. A possible type declaration could be

```
type cell = record
              element : elemtype;
              next : 0..m
            end;
     list = record
              elements : array [1..m] of cell;
              first, empty : 0..m
            end;
```

A similar kind of simulation of pointers will be discussed in the next chapter on implementation of trees.

To consolidate the ideas of abstract date type and the linear ADTs that we have studied, we now look at a fairly large application using these and discuss the development of the final program.

3.4 Case Study: Evaluation of Arithmetic Expressions

This case study involves developing a program for reading, storing and evaluating arithmetic expressions. Expressions are all presented to the program in infix notation with brackets to eliminate ambiguities. The arithmetic operators are given precedences as follows:

$$\begin{array}{c} * / \\ + - \end{array} \Big\downarrow \quad \text{lower precedence}$$

A sequence of operators of equal precedence (with no bracketing) should be evaluated from left to right. For example

A / B * 20 = (A / B) * 20
15 * B + F − 10289 / D ≡ ((15 * B) + F) − (10289 / D)

It may be assumed that all the variables used in the expressions are capital letters 'A' . . . 'Z', and only positive integers are allowed to appear in the expressions (in addition to variables). Thus the operator '/' should be taken as equivalent to **div** in Pascal.

The statements in the input are of two forms:

(1)   ⟨Identifier⟩ = ⟨Expression⟩
(2)   ⟨Expression⟩

Type (1) statements define identifiers whereas statements of type (2) indicate that an expression is to be evaluated. If, in a type (1) command, ⟨Expression⟩ is empty, the ⟨Identifier⟩ should be undefined. The definitions can be stored in Reverse Polish notation. While evaluating an expression, if there are variables which are either not defined or cannot be fully evaluated, the output should be given in Reverse Polish (post-fix) notation.

An example of input and output would be like this:

A = B + C − D / F
O.K.

C = B + 3
O.K.

2 * A + 3
2 B B 3 + + D F / − * 3 +

B + 30
O.K.

C
33

B + 2 * C + 10
106

B = 25
O.K.

C + 2
30

. . . .
. . . .
. . . .
$

The underlined text are responses of the program to the input statements.

Development of the Program

Already in this chapter we have studied how to evaluate expressions in Reverse Polish. The major data structure we need is for storing the definition of identifiers. Such a data structure can be imagined as a set where each member is the definition of an identifier. We shall discuss implementation of sets later in the book. For this problem, since there can be a maximum of 26 identifiers, we can define a table containing these expressions as an array:

```
table : array['A'..'Z'] of texpr;
```

Each element is a *table expr*ession (texpr). An expression in Reverse Polish is naturally a sequence of symbols and therefore a list can be used to represent it. A table expression can, of course, be empty if there is no definition for a given identifier. Therefore, texpr can be defined as

```
texpr = record
          case defined : boolean of
                true : (seq : LIST);     (* a list of symbols *)
                false : ()
          end;
```

LIST refers to the ADT LIST of symbols. We need to define a symbol (symbol substitutes the type elemtype). A symbol is either an identifier, an operator, a number, an equal sign or a closed or open bracket. An operator can be an addition (plus), a subtraction (minus), a multiplication (mult) or a division (divide) operator. Therefore, the following simple data types can be defined:

```
type
  arop = (plus, minus, mult, divide);   (* arithmetic operator *)
  symtype = (int, op, id, null, equal, openB, closedB);
   (* This is the types of symbols. The type null
      is used to refer to an empty symbol *)
  symbol = record
             case kind : symtype of
                   int : (value : integer);
                   op : (operator : arop);
                   id : (identifier : 'A'..'Z');
                 null, equal, openB, closedB : ()
           end;
```

With these types, we can easily define and implement the ADT's *stack of symbol* and *list of symbol*. Then, an expression is simply a list of symbols. The overall structure of the program can thus be defined as

```
repeat
    read an expression and transform it to reverse polish
    if not end of input then
        if it is a definition then add it to the table
        else evaluate the expression and print the result
until end of input;
```

Using this framework we can give the following outline of the program:

```
program case_study(input, output);
  type
    arop = (plus, minus, mult, divide);
    symtype = (int, op, id, null, equal, OpenB, CloseB);
    symbol = record
               case kind : symtype of
                 int : (value : integer);
                 op  : (operator : arop);
                 id  : (identifier : 'A'..'Z');
                 null, equal, OpenB, CloseB : ()
             end;
    expr = LIST;
    texpr = record                      (* Table expression *)
              case defined : boolean of
                true  : (seq : LIST);
                false : ()
            end;
  var
    expbuff : expr;
    S : STACK;
    table : array['A'..'Z'] of texpr; (* The table of expressions *)
    end_of_input : boolean;           (* To indicate the end of input *)

(* The Abstract Data Type STACK OF SYMBOL *)

(* The Abstract Data Type LIST OF SYMBOL *)

procedure initialiseTable;
  (* This procedure empties the table of expressions *)

procedure read_a_line(ebuf : expr);
  (* This procedure reads a line of input and stores the content
     in the LIST ebuf. If The expression is ill-formed, it is
     rejected and another line is attempted. if the first
     character read is a '$', the end_of_input is set to 'true' *)

function definition(ebuf : expr) : boolean;
  (* This function returns 'true' if ebuf is a definition
     of an expression and 'false' if it is an expression to
     be evaluated *)

procedure add_to_table(ebuf : expr);
  (* This procedure adds the definition of expression ebuf to the
     table.if the new definition causes the expression to be
     cyclic, the operation is rejected *)

procedure evaluate (ebuf : expr);
  (* This procedure evaluates the expression ebuf *)

(* Main Program *)
begin
(* initialise Data Types *)
  initialise(S);
  initialiseTable;
```

```
repeat
 read_a_line(expbuff);  (* read a line in reverse polisyh *)
 if not end_of_input then
  begin
   if definition(expbuff) then

      add_to_table(expbuff)
     else
       evaluate(expbuff)
    end
   until end_of_input;

 end.  (* of program *)
```

The implementations of stacks and lists were discussed earlier in the chapter. For both, we use the array implementation. This is because for lists we only need operations initialise, insert_at_end, retrieve, empty, full, next, previous, first and last. Therefore, a simple array implementation is sufficient to give efficient processing (all the operations need a time of O(1)). Note that a new operation insert_at_end is introduced to abstract the particular mode of operation of this program. Also, the operations top and retrieve are given as procedures with variable parameters rather than functions. This is because the type symbol is a record and Pascal functions are only allowed to return a simple type (integer, sub-range or an enumerate type) or a pointer.

Read_a_line
This procedure reads a line and transforms it to Reverse Polish. An extension to the algorithm we have described in this chapter is that '=' is treated like any other symbol with a low precedence.

$$
\begin{array}{l}
* \,/ \\
+ \,- \\
(\\
= \\
\text{null}
\end{array}
\quad \text{decreasing precedence}
$$

For instance, A = B * C / (D * 10) will be transformed to

    A B C * D 10 * / =

Thus, the equal symbol, if there is one, should come last. This procedure skips spaces (procedure skipspaces) and non-valid characters. The procedure getsymbol is defined, which gets the next symbol in the input line. The read_a_line then uses the algorithm developed in this chapter to transform an expression to RP. There is one exception to this: since there are no ';' at the end of each expression, the operators on top of the stack should be printed out and popped accordingly (at the end).

Definition

An expression in RP is a definition if the last symbol is an equal symbol.

Add_to_table

Given an expression, a check must be made that it is not cyclic. If this is so, it can be added to the table. The procedure cyclic(ch, ebuf, cycles) returns cycles = true if the expression ebuf contains the identifier ch (in a recursive manner). Before calling cyclic, cycles is set to false and the procedure cyclic will reset it to true if ebuf references 'ch' directly or recursively.

Evaluate

This is a recursive procedure to evaluate an expression ebuf. The algorithm is exactly that of this chapter with one difference. While scanning the expression, if an identifier is encountered and is defined (that is, if it has a definition in the table), then a recursive call is made to evaluate the expression corresponding to that identifier. If the identifier is undefined, however, it is just treated as a number and is pushed into the stack.

Finally, a procedure is used to print out the result of an evaluation. The evaluate procedure leaves the answer in the stack. (It may be an expression if some identifiers are undefined; otherwise a number will be on top of the stack.) Since the result is in the stack, a sequence of top and pop operations will yield the inverse of the result. Thus an implicit stack (via recursion) can be used to print the result in the correct order. The procedure print_result first calls itself recursively and then prints the current symbol. The entire program and some input/output are given below.

```pascal
program case_study(input, output);
  label
    999; (* to exit the program *)
  const
    max = 20;      (* Maximum no. of elements in a stack or a list *)
    maxplus1 = 21; (* max + 1 *)
  type
    arop = (mult, divide, plus, minus);
    symtype = (int, op, id, OpenB, CloseB, equal, null);
    symbol = record
               case kind : symtype of
               int : (value : integer);
               op : (operator : arop);
               id : (identifier : 'A'..'Z');
               null, equal, OpenB, CloseB : ()
             end;
    STACK = record
              ttop : 0..max;
              elements : array[1..max] of symbol
            end;
    POSITION = 1..maxplus1;
    LIST = record
             eol : 0..max;
```

```
                    elements : array[1..max] of symbol;
                end;
      expr = LIST;
      texpr = record
                  case defined : boolean of
                      true : (seq : LIST);
                      false : ()
                end;
var
   expbuff : expr;
   S : STACK;
   table : array['A'..'Z'] of texpr;
   end_of_input : boolean;
   emptyexpr : expr;
   nullsym : symbol;
   i : integer;
   p : position;

procedure error (errorno : integer); forward;
   (* This procedure is used to exit the program
      when an error has occurred *)


(* The Abstract Data Type STACK OF SYMBOL and its implementation *)

function empty_stk (s : stack) : boolean;
   begin
      empty_stk := (s.ttop = 0)
   end; (* of empty_stk *)

function full_stk (s : stack) : boolean;
   begin
      full_stk := (s.ttop = max)
   end; (* of full_stk *)

procedure stk_initialise (var s : stack);
   var i : integer;

   begin
      s.ttop := 0;
      for i := 1 to max do s.elements[i].kind := null
   end; (* of stk_initialise *)

procedure push (var s : stack; a : symbol);
   begin
      if full_stk(s) then
         error(1)
      else
         begin
            s.ttop := s.ttop + 1;
            s.elements[s.ttop] := a
         end
   end; (* of push *)

procedure pop (var s : stack);
   begin
      if empty_stk(s) then
         error(2)
```

```
    else
      s.ttop := s.ttop - 1
  end; (* of pop *)

procedure top (s : stack; var sy : symbol);
  begin
    if empty_stk(s) then
      error(2)
    else
      sy := s.elements[s.ttop]
  end; (* of top *)


(* The Abstract Data Type LIST OF SYMBOL and its implementation *)

function full_list (L : list) : boolean;
  begin
    full_list := (L.eol = max)
  end;

function empty_list (L : list) : boolean;
  begin
    empty_list := (L.eol = 0)
  end; (* of empty_list *)

function first (L : list) : position;
  begin
    first := 1
  end; (* of first *)

function last (L : list) : position;
  begin
    last := L.eol + 1
  end; (* of last *)

function next (p : position; L : list) : position;
  begin
    if p = last(L) then
      error(5)
    else

      next := p + 1
  end; (* of next *)

function previous (p : position; L : list) : position;
  begin
    if p = first(L) then
      error(6)
    else
      previous := p - 1
  end; (* of previous *)

procedure list_initialise (var L : list);
  var i : integer;
  begin
    L.eol := 0;
    for i := 1 to max do L.elements[i].kind := null
  end; (* of list_initialise *)
```

```
procedure insert_at_end (var L : list; a : symbol);
  begin
    if full_list(L) then
      error(3)
    else
      begin
        L.eol := L.eol + 1;
        L.elements[L.eol] := a
      end
  end; (* of insert_at_end *)

procedure insert(var L : list; a : symbol; p : position);
  var i : position;
  begin
    if p = max then error(3)
    else if (p < 1) or (p > L.eol) then error(7)
        else
          begin
            for i := L.eol downto p do
              L.elements[i+1] := L.elements[i];
            L.eol := L.eol + 1;
            L.elements[p] := a
          end
  end; (* of insert *)

procedure delete(var L : list; p : position);
  var i : position;
  begin
    if (p < 1) or (p > L.eol) then error(7)
    else
      begin
        for i := p to (L.eol - 1) do
          L.elements[i] := L.elements[i+1];
        L.eol := L.eol - 1
      end
  end; (* of delete *)

procedure retrieve(L : list; p : position; var s : symbol);
  begin
    if p = last(L) then
      error(7)
    else
      s := L.elements[p]
  end; (* of retrieve *)

procedure error;
  begin
    case errorno of
      1 : writeln('stack is full');
      2 : writeln('stack is empty');
      3 : writeln('list is full');
      4 : writeln('list is empty');
      5 : writeln('no next position');
      6 : writeln('no previous position');
      7 : writeln('No such position')
    end;
    goto 999
  end; (* of error *)
```

```
(* The procedures and functions of this program *)

procedure initialiseTable;
  (* The procedure empties the table of expressions *)
  var
    ch : char;
  begin
    for ch := 'A' to 'Z' do
      table[ch].defined := false
  end; (* of initialise_table *)

procedure read_a_line (var ebuf : expr);
  (* This procedure reads a line of input and stores the
      content in the LIST ebuf. If the expression is ill-formed,
      it is rejected and another line is attempted. If the
      first character read is a '$', the end_of_input is set
      to 'true' *)
  var
    eoline, eoi : boolean;
    sym, temp : symbol;
    stk : stack;
    oper : arop;

  function less_or_equal (s1, s2 : symbol) : boolean;
    var
      p1, p2 : integer;
    function prec (s : symbol) : integer;
      begin
        case s.kind of
          op : case s.operator of
                  mult, divide : prec := 1000;
                  minus, plus : prec := 500
                end;
            OpenB : prec := 300;
            equal : prec := 200;
            null : prec := 100
        end
      end; (* of prec *)
    begin
      less_or_equal := (prec(s1) <= prec(s2))
    end; (* of less_or_equal *)

procedure getsymbol (var sym : symbol; var eoi, eoline : boolean);
    (* Read the next symbol, if '$' found, eoi is set to true,
        if end of line is encountered, eoline is set to true and the
        line i skipped *)

  var
    ch : char;
    no : integer;
  procedure skipspaces;
    begin
      while (input^ = ' ') and not eoln do
        get(input)
    end; (* of skipspaces *)

  begin
    skipspaces;
```

```
    if eoln then
      begin
        readln; eoline := true
      end
    else
      begin
        read(ch);
        if ch = '$' then
          eoi := true
        else
          begin              (* non-valid characters are ignored *)
            if not (ch in ['0'..'9','A'..'Z','+','-',
                                     '*','/','(',')','=']) 
            then getsymbol(sym, eoi, eoline) else
            if ch in ['A'..'Z'] then
              begin
                sym.kind := id;
                sym.identifier := ch
              end
            else
              case ch of
                '(' : sym.kind := OpenB;
                ')': sym.kind := CloseB;
                '+', '-', '*', '/' :
                  begin
                    sym.kind := op;
                    case ch of
                      '+': oper := plus;
                      '-' : oper := minus;
                      '*' : oper := mult;
                      '/' : oper := divide
                    end;
                    sym.operator := oper
                  end;
                '0', '1', '2', '3', '4', '5', '6', '7', '8', '9' :
                  begin (* read a number and convert to integer *)
                    no := ord(ch) - ord('0');
                    while input^ in ['0'..'9'] do
                      begin
                        read(ch);
                        no := no * 10 + (ord(ch) - ord('0'))
                      end;
                    sym.kind := int;
                    sym.value := no

                  end;
                '=' : sym.kind := equal
              end (* of case *)
          end;
        end;
    end; (* of getsymbol *)

(* read_a_line *)
begin
  writeln('_____');
  list_initialise(ebuf);
  stk_initialise(stk);
  push(stk, nullsym); (* The bottom of stack *)
```

```
   eoline := false; eoi := false;
while not eoline do
  begin
     getsymbol(sym, eoi, eoline);
     if eoi then
       begin
         end_of_input := true; eoline := true
       end
     else if eoline then
             begin
                (* if an empty line is read,
                   a new line should be read *)
                if empty_list(ebuf) then
                  read_a_line(ebuf)
             end
          else
             begin
                (* Construct Reverse Polish Expression *)

(* Error Checking is not done in this implementation.
   These include :
   1) open bracket with no close bracket e.g.  A+(B + C
   2) close bracket with no open bracket e.g.  A + C + D)
   3) case sym of
         OP : previous sym should be ID or INT or
                  next sym should be ( or ID or INT
      INT,ID : previous sym should be OP or ( or =
         (  : previous sym should be OP or = or (
         )  : previous sym should be INT or ID or )
         =  : previous sym should be ID and the only ID so far *)

                case sym.kind of
                   int, id : insert_at_end(ebuf, sym);
                   OpenB, equal : push(stk, sym);
                   CloseB :
                     begin
                        top(stk, sym);
                        while sym.kind <> OpenB do
                          begin
                             insert_at_end(ebuf, sym);
                             pop(stk);
                             top(stk, sym)
                          end;
                        pop(stk) (* To remove '(' *)
                     end;
                   op :
                     begin
                        top(stk, temp);
                        while less_or_equal(sym, temp) do
                          begin
                             insert_at_end(ebuf, temp);
                             pop(stk); top(stk, temp)
                          end;
                        push(stk, sym)
                     end;
                end;  (* of case *)
             end;
```

```
      end;
    top(stk, sym);
    while sym.kind <> null do (* to empty the stack *)
      begin
        insert_at_end(ebuf, sym);
        pop(stk);
        top(stk, sym);
      end;
  end;  (* of Read_a_line *)

function definition (ebuf : expr) : boolean;
  (* This function returns 'true' if the ebuf is a definition
     of an expression and 'false' if it is an expression to
     be evaluated *)
  var
    sym : symbol;
  begin
  (* ebuf is a definition of an expression if the
     last element of the expr in reverse polish
     is an 'equal' symbol *)
  retrieve(ebuf, previous(last(ebuf), ebuf), sym);
  definition := (sym.kind = equal)
  end; (* of defnition *)

procedure add_to_table (ebuf : expr);
  (* This procedure adds the definition of expression ebuf to the
     table. If the new definition causes the expression to be
     cyclic, the operation is rejected *)
  var
    p : position;
    letter : 'A'..'Z';
    sym, s : symbol;
    temp : expr;
    cycles : boolean;

  procedure cyclic (ch : char; ebuf : expr; var cycles : boolean);
    var
      p : position;
      sym : symbol;

    begin
      p := first(ebuf);
      while (p <> last(ebuf)) and not cycles do
        begin
          retrieve(ebuf, p, sym);
          if sym.kind = id then
            if sym.identifier = ch then
              cycles := true
            else if table[sym.identifier].defined then
              cyclic(ch, table[sym.identifier].seq, cycles);
          p := next(p, ebuf)
        end; (* of while *)
    end; (* of cyclic *)

  begin
    p := first(ebuf);
    retrieve(ebuf, p, sym);
```

```
letter := sym.identifier;
(* This is the letter being defined *)

(* CHECK FOR CYCLIC DEFINITIONS *)

list_initialise(temp);
(* temp holds the definition of letter *)
p := next(first(ebuf), ebuf);
while p <> previous(last(ebuf), ebuf) do
  begin
    retrieve(ebuf, p, s);
    insert_at_end(temp, s);
    p := next(p, ebuf)
  end;
cycles := false;
cyclic(letter, temp, cycles);
if cycles then
  writeln('Cyclic definition')
else
  begin
    (* The first and last symbols of
       the ebuf must be discarded, *)
    (* i.e. 'letter' and '=' *)
    if next(first(ebuf), ebuf)
                  = previous(last(ebuf), ebuf) then
      table[letter].defined := false
      (* This occurs when an identifier is un-defined *)
    else
      begin
        table[letter].defined := true;
        list_initialise(table[letter].seq);
        p := next(first(ebuf), ebuf);
        while p <> previous(last(ebuf), ebuf) do
          begin
            retrieve(ebuf, p, s);
            insert_at_end(table[letter].seq, s);
            p := next(p, ebuf)
          end
      end;
    if not table[letter].defined then
      writeln('No expression. O.K.')
    else
      writeln('O.K.')
  end
end; (* of Add_to_table *)

procedure evaluate(ebuf : expr);
  (* This procedure evaluates the expression ebuf *)
  var
    p : position;
    sym, s1, s2 : symbol;
    eb : expr;
        begin
          p := first(ebuf);
          while p <> last(ebuf) do
            begin
              retrieve(ebuf, p, sym);
```

```
          case sym.kind of
            int : push(s, sym);
            id :
              begin
                if table[sym.identifier].defined then
                  begin
                    eb := table[sym.identifier].seq;
                    evaluate(eb)
                  end
                else
                    push(s,sym)
              end;
            op :
              begin
                top(s, s1);
                pop(s);
                top(s, s2);
                pop(s);
                if (s1.kind <> int) or (s2.kind <> int) then
                  begin
                    push(s, s2);
                    push(s, s1);
                    push(s, sym)
                  end
                else
                  begin
                    case sym.operator of
                      plus : begin
                               s1.value := s1.value + s2.value;
                               push(s, s1)
                             end;
                      minus : begin
                                s1.value := s2.value - s1.value;
                                push(s, s1)
                              end;
                      mult : begin
                               s1.value := s1.value * s2.value;
                               push(s, s1)
                             end;
                      divide : begin
                                 s1.value := s2.value div s1.value;
                                 push(s, s1)
                               end;
                    end; (* of case sym.operator *)
                  end;
              end;
          end; (* of case sym.kind *)
          p := next(p,ebuf);
        end; (* of while loop *)
   end; (* of Evaluate *)

procedure print_result;
   var
     sym : symbol;
   begin
     if not empty_stk(s) then
       begin
```

```
            top(s, sym);
            pop(s);
            print_result;
            case sym.kind of
               int : write(sym.value : 3,' ');
               id: write(sym.identifier,' ');
               op: case sym.operator of
                      mult : write('X ');
                      divide : write('/ ');
                      minus : write('- ');
                      plus: write('+ ')
                   end;
            end
         end
      end; (* of print_result *)

(* Main Program *)
begin
  (* Initialise Data Types *)
  stk_initialise(s);
  list_initialise(expbuff);
  list_initialise(emptyexpr); (* an empty expression *)
  nullsym.kind := null;        (* a null symboll *)
  initialiseTable;
  writeln('Type expressions one at a time on separate lines.');

  repeat
    read_a_line(expbuff); (* read a line in reverse polish *)
    if not end_of_input then
      begin
        if definition(expbuff) then
          add_to_table(expbuff)
        else
          begin
            Evaluate(expbuff);
            (* Print the result *)
            writeln('The result is :-');
            print_result;
            writeln
          end
      end
    until end_of_input;

  999 :
  writeln('End of Execution.')
end. (* of program *)
```

Here is a sample of the input/output of the program case_study:

```
Type expressions one at a time on separate lines
```

```
A=B*C
O.K.
```

```
B=35*(((A/2)))
Cyclic definition.
```

```
B=123
O.K.
```

```
A*2
The result is :-
123 C * 2 *
```

```
B=
No expression. O.K.
```

```
(A/120)+(12*23+6)
The result is :-
B C *  120 /  282 +
```

```
109*45+5
The result is :-
4910
```

```
D=E+34*G
O.K.
```

```
100*(34+D)
The result is :-
100  34 E 34 G * + + *
```

```
$End of Execution.
```

Exercises

3.1. Write a *recursive* procedure to read a sequence of characters and write it in reverse order (note that an implicit stack is used).

3.2. A stack S is used in the following manner:

> initialise(S);
> (∗ sequence of statements using the stack S ∗)
> initialise(S);
> (∗ sequence of statements using the stack ∗)
> .
> .
> .

What is the side-effect of this sequence of statements and how can it be remedied? (Consider pointer-implementation of stacks.)

3.3. Using a list, implement an ADT's stack and queue.

3.4. Implement the operation Assign for all ADTs in this chapter. Also, define and implement a procedure

> **procedure** clear (**var** a : ADT);

which would empty the structure 'a' for all the linear abstract data types of this chapter.

3.5. (a) The *Tower of Hanoi* problem can be solved as follows:

```
(*  1 ≤ i,j,k ≤ 3 *)
procedure Hanoi(n,i,j,k);
   (* moves discs from pole i to pole k using pole j *)
   begin
   if n = 1 then move the top disc from pole i to k
   else begin
        Hanoi(n-1,i,k,j);
        (* move the top disc from i to k *)
        Hanoi(n-1,j,i,k)
      end
   end;
```

Rewrite this algorithm without using recursion. Note that there is a tail recursion.

(A non-recursive Hanoi algorithm can be developed which does not use a stack at all. See *ACM SIGPLAN*, Vol. 19, No. 9, September 1984.)

3.6. Verify that all stack and queue operations need a time of O(1).

3.7. In the evaluation of expressions, modify the procedure such that it produces (compiles) code for a simple machine with instructions:

> ⟨OP⟩ ⟨Memory⟩ ⟨R⟩   (∗ ⟨R⟩ := ⟨R⟩ ⟨OP⟩ ⟨Memory⟩ ∗)
> MOVE ⟨Memory⟩ ⟨R⟩   (∗ ⟨R⟩ := ⟨Content of Memory⟩ ∗)

STORE ⟨R⟩ ⟨Memory⟩   (∗ Content of Memory := ⟨R⟩ ∗)

where ⟨OP⟩ can be ADD, SUB, MULT and DIVIDE; ⟨memory⟩ is the address of a memory location (for example, a, b etc.), and ⟨R⟩ is a register of the computer.

3.8. To implement *last* efficiently using the pointer implementation a *ring* structure can be used:

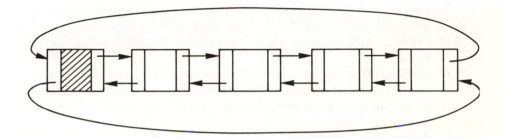

Write the list operations for the above structure.

How can a singly-linked ring structure be used for implementing lists?

3.9. Complete the simple array implementation of lists.

3.10. What is the effect of the following fragment:

        initialise(L1); (∗ L1 and L2 are lists ∗)
        initialise(L2);
        (∗ insert 10 items into list L1 ∗)
        L2 := L1;
        delete(L2,first(L2));
        (∗ print the entire list L1 ∗)

Assume that a pointer implementation of lists is used.

3.11. Define and implement the operations (using pointers):

        procedure assign(L1: list; **var** L2 : list);
            (∗ L2 := L1 ∗)
        function equal(L1,L2 : lists) : boolean;
            (∗ equal is true if L1 and L2 are exactly similar. ∗)
        Does equal := (L1 = L2) do?

3.12.

```
procedure X (L :list; a : elemtype);
  var p : position;
  begin
    p := last(L);
    insert(L,a,p)
  end;
```

```
begin  (* main program *)
  initialise(L); (* L is a variable of type list *)
  P := first(L);
  insert(L,x,p);
  (*     * 1     *)
  X(L,a);
  (*     * 2     *)
end;
```

> What are the values of L before and after the call X(L, a)? Note that L is a value parameter of X. Consider both array and pointer implementations of list.

3.13. Define an ADT 'sequence' which would abstract the 'file' data structure of Pascal.

3.14 (Difficult) How can the *value* field of the texpr of the case study be used to make the evaluation process faster? (Use a non-recursive procedure evaluation and aim at evaluating an identifier not more than once if necessary.)

3.15. Augment the program of case study, section 3.4, to include full error checking.

Hint: (i) Most errors can be trapped by maintaining the 'current' and 'previous' symbols in the expression.

(ii) For checking mismatched brackets, either (a) use a 'counter' to represent

> (number of open brackets) – (number of closed brackets)

or (b) check the state of the stack when a close bracket is encountered, or when the whole expression is read.

3.16. In the evaluation of expressions, how can arithmetic overflow (or underflow) be detected? (Assume that the underlying computer does not detect these.)

3.17. A roundabout contains five roads (see diagram). The arrows show the direction of traffic. A program is to be written to simulate the behaviour of the traffic. For each car approaching the roundabout it is known which turn it is intending to take. Discuss the abstract data types and their implementations in terms of other ADTs and/or data structures of Pascal to represent the data in this application. Clearly state any assumptions you make. Briefly comment on the operation of such a program (with reference to your ADTs) as a car is about to enter the roundabout until it leaves the roundabout at the intended exit.

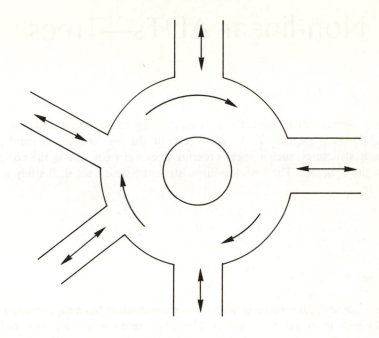

Bibliographic Notes and Further Reading

Knuth's *The Art of Computer Programming* (Knuth, 1973) is a rich source of material on lists, stacks and queues. Implementations of multiple stacks are given which make efficient use of storage. For lists, implementations are given which optimise the 'amortised' time complexity of the insert and delete operations. This refers to the time needed to perform k insert and delete operations on a list (also see the notes at the end of chapter 1). The file data structure of Pascal can be thought of as an implementation of lists, with the major difference that the data type is persistent (for external files) — that is, it exists even after the program has terminated. Files are abstractions of simpler lists, where operation insert only applies to the end of the list and retrieval is strictly sequential (exercise **3.13**).

Knuth, D. E. (1973). *The Art of Computer Programming. Volume 1: Fundamental Algorithms*, 2nd edition, Addison-Wesley, Reading, Massachusetts.

4 Non-linear ADTs—Trees

In linear abstract data types, there exists exactly one previous and one next element for each element of the ADT (except the first and last elements). In non-linear structures, such a linear ordering does not exist among the components of the structure. The first non-linear structure which we shall study is the ADT tree.

4.1 Trees

Trees are hierarchical structures where every component has a super-component and may have many sub-components. These two terms intuitively represent the previous and the next elements of an individual element. Examples of trees are numerous in almost all fields of study. For example, the structure of this book can be represented as a tree (figure 4.1). The book is broken down into a few parts and chapters. Each chapter in turn is broken down into many sections, and so on. A second example is the representation of the statements of a high-level language. For instance, the structure of a simple **if** statement of Pascal can be represented as in figure 4.2.

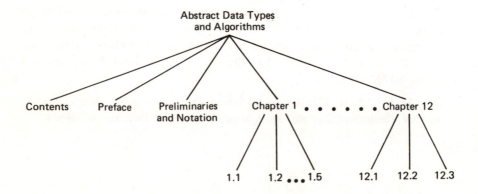

Figure 4.1 Structure of this book.

if C = 10 **then** A := 10
else A := 10 + B;

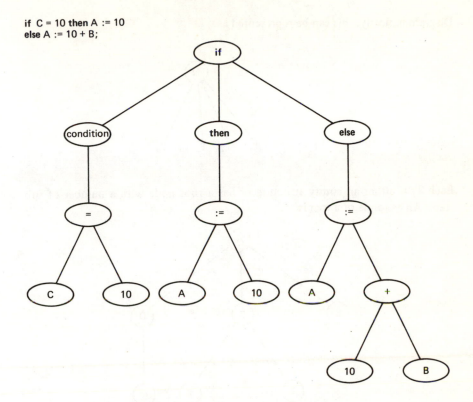

Figure 4.2   Structure of an **if** statement

Structures such as those of figure 4.2 are constructed by compilers which are then used to produce executable code for a given program.

Definitions

A tree is a recursive abstract data type which is a collection of zero or more *nodes* each containing an element of type *elemtype* with the following hierarchical structure:

either (a) The empty structure (with no nodes)
or   (b) A node N containing element E with a finite number (zero or more) of associated disjoint tree structures (that is, they have no common nodes). These trees are known as *subtrees* of N. N is called the *root* of the tree.

Diagrammatically, this can be represented as:

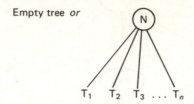

Empty tree *or*

Each T_i is either an empty structure or has a root node with a number of sub-trees. An example of a tree is

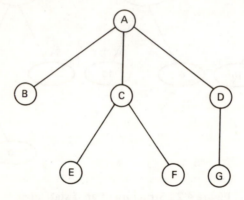

Nodes with no subtrees are called *terminal* or *leaf* nodes. Nodes B, E, F and G are leaf nodes. Other nodes are *internal* nodes. N is the *parent* of the roots of trees $T_1, T_2, \ldots, T_n$. In the example above: A is the parent of nodes B, C and D; C is the parent of nodes E and F; and finally D is the parent of node G. The root node of each T_i is a *child* of N. The root nodes of each T_i are *siblings* of one another. (Note that this terminology is derived from family trees which represent the relationships between parents and their children.) Nodes B, C and D are siblings, and so are nodes E and F.

The children of a node N can be ordered or unordered. The resulting trees are *ordered* and *unordered* respectively. Thus the following two ordered trees are not equivalent:

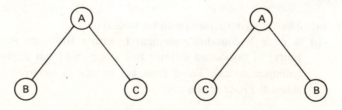

In order to construct and manipulate trees, we define an abstract data type tree with a number of operations. Since the number of children of each node is not fixed, an effective way to manipulate trees is to be able to access the left-most child and then successively access all the siblings of that child for a given node. Therefore, we define a tree as a set of nodes with a hierarchical structure. In the following, we assume that one element of type node is designated as a null node. The following set of operations can be defined on ordered trees:

procedure initialise (**var** T : tree);
    (∗ This procedure returns an empty tree ∗)

function empty (T : tree) : boolean;
    (∗ If T is empty a true value is returned, otherwise a false
    value is returned ∗)

function root (T : tree) : node;
    (∗ If T is empty, this function returns a null node. For non-empty trees T,
    the root node of the tree T will be returned ∗)

function retrieve (N : node) : elemtype;
    (∗ This function returns the element stored in node N ∗)

function parent (N : node) : node;
    (∗ This function returns the parent node of N.
    If N has no parent, a null node will be returned ∗)

function left_most_child (N : node) : node;
    (∗ This function returns the left-most child of the node N.
    If there are no children, a null node will be returned ∗)

function right_sibling (N : node) : node;
    (∗ If node N has a sibling node, it will be returned as the value of this
    function. Otherwise, a null node will be returned ∗)

function null_node (N : node) : boolean;
    (∗ This function returns a true value if and only if N is a null node ∗)

function cons_tree$_n$ (E : elemtype; $T_1, T_2, \ldots, T_n$: tree) : tree;
    (∗ This function constructs a tree with a node N as its root containing
    element E and the trees $T_1, T_2, \ldots, T_n$ as its children, in that order.
    The trees $T_1, T_2, \ldots, T_n$ should not be used again after this operation
    to construct other trees. If copies of these subtrees are required, they
    can be constructed using the assign(T, T′) operation. This is meant for
    space-efficiency so that this function need not copy these subtrees in
    order to make a new tree ∗)

procedure insert_child (P : node; E : elemtype);
    (∗ This procedure is defined for constructing trees in a top-down manner.
    A node containing element E will be created and it will be made the
    right-most child of node P ∗)

procedure delete (**var** T : tree);
    (∗ This procedure deletes all the nodes of the tree T, and T will be made
      an empty tree. The tree T should not be a subtree of another tree ∗)

procedure assign (T1 : tree; **var** T2 : tree);
    (∗ This procedure copies the entire tree T1 and asigns it to T2 ∗)

Note that, in the above definitions, two data types, namely nodes and trees,
are introduced. For instance, to construct the following tree

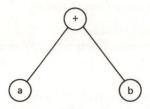

the necessary sequence of operations are (assuming that elemtype is defined as
char):

    T1 := cons_tree$_0$ ('a');   (∗ to construct tree $\;a\;$ ∗)
    T2 := cons_tree$_0$ ('b');
    T   := cons_tree$_2$ ('+', T1, T2);   (∗ to construct tree $\;b\;$ ∗)

                                        (∗ to construct $\;+\;$ ∗)

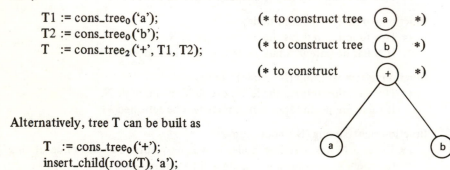

Alternatively, tree T can be built as

    T   := cons_tree$_0$ ('+');
    insert_child(root(T), 'a');
    insert_child(root(T), 'b');

An alternative way to define the ADT tree is to regard a tree as either an empty
structure or a root node and a *list* of children trees (see exercise **4.31**).

Processing Tree Structures

Processing trees usually requires processing individual nodes of trees. To be able
to visit and process all the nodes of a tree, there are primarily three systematic
ways. These are known as *tree walkings* or *tree traversals*. In a tree with the root
N and subtrees $T_1, T_2, \ldots, T_n$ we define:

    pre-order traversal: visit the root node N, then
                         visit all the nodes in T_1 in pre-order, then
                         visit all the nodes in T_2 in pre-order,

and so on until all the nodes in T_n are visited
in pre-order.

post-order traversal: visit the nodes of $T_1, T_2, \ldots, T_n$
in post-order (in that order); then
visit the root node.

in-order traversal: visit all the nodes of T_1 in in-order then
visit the root node; then visit the nodes of
$T_2, T_3, \ldots, T_n$ in in-order (in that order).

Thus, for the tree of figure 4.2 these traversals are

pre-order: **if** condition = C 10 **then** := A 10 **else** := A + 10 B
post-order: C 10 = condition A 10 := **then** A 10 B + := **else if**
in-order: C = 10 condition **if** A := 10 **then** A := 10 + B **else**

Using the operations on trees, these traversals can be written according to the
above definitions. For instance, for the pre-order traversal, the following algorithm can be used:

```
procedure pre_order(N : node);
  var temp : node;
  begin
    (* process node N *)
    temp := left_most_child(N);
    while not null_node(temp) do
      begin
        pre_order(temp);    (* the recursive step *)
        temp := right_sibling(temp)
      end
  end;  (* of pre_order *)
```

The call pre_order(root(T)) will perform a pre-order traversal of the nodes of
tree T. This procedure first visits node N, and then visits the other nodes by
traversing all the children nodes of N in pre-order recursively.

Implementation of Trees

1. Pointer implementations

Each node of a tree can be represented as a record consisting of an element and
two pointers to the left-most and right-sibling trees. If necessary an extra pointer
can be used to indicate the parent of a node. The ADT tree, therefore, can be
defined as a pointer to the root node of a tree.

```
type node = ↑node_cell
     node_cell = record
                     element : elmtype,
                     left_most, right_sib : node
                 end;
     tree = node;
```

Figure 4.3 shows a simple example.

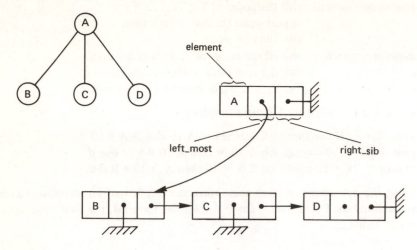

Figure 4.3   Pointer representation of trees.

An empty tree is represented by a *nil* pointer. The following program shows the implementation of the function right_sibling and the procedure assign.

Implementation of Right_sibling and Assign

```
function right_sibling (N : node) : node;
  begin
    if null_node(N) then error
    else right_sibling := N↑.right_sib
  end;

procedure assign(T1 : tree; var T2 : tree);
  var P, Q : tree;
  begin
    if T1 = nil then T2 := nil
    else
      begin
        assign(T1↑.left_most, P);
        assign(T1↑.right_sib, Q);
        new(T2);
        T2↑.element := T1↑.element;
        T2↑.left_most := P;
        T2↑.right_sib := Q
      end
  end;
```

All other operations except 'parent' function are implemented easily. If this function is required frequently, a first solution is to use an extra pointer in each node to point to its parent. However, this increases the extra storage needed by $O(n)$. An alternative solution is to use the *nil* right-sibling pointers to point to the parent node. Naturally, the root node has no parent and its right-sibling pointer will be set to nil. For instance, the tree of figure 4.3 would be represented as

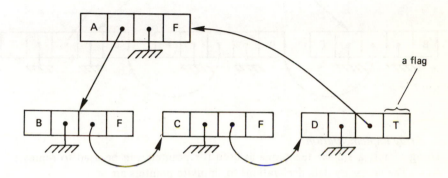

An extra flag is needed to show that end of a right-sibling list is reached. For instance, if flag = true then a parent link is expected. Using this structure the parent function can be written:

```
function parent (N : node) : node;
  var T : node;
  begin
    T := N;
    while not (T↑.flag) and (T↑.right_sib <> nil) do
        T := T↑.right_sib;
    parent := T↑.right_sib
  end;  (* of parent *)
```

This usage of the right-sib pointers is further discussed later (see threaded-trees, section 4.3). Note that the above implementation affects the other operations on trees.

The implementation of the insert-child operation requires a search through the right-sib pointers of children of a node to find the right place to add a new node. If this operation is executed frequently, every node may have an extra pointer which would locate the last child of that node. An example of this is given below:

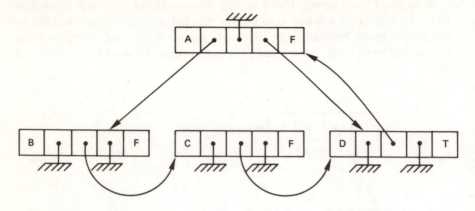

2. Array implementations

Using arrays, a similar technique as used for pointers can be used to represent trees. The necessary data declarations to simulate pointers are

```
const N =    ;  (* max number of nodes *)
type tree = 0..N;       (* 0 indicates an empty tree *)
     node_cell = record
                      element : elemtype;
                      left_most, right_sib : tree
                    end;
     node = 0..N; (* 0 indicates a null node *)
var table : array [1..N] of node_cell;
    T : tree;
```

	element	left_most	right_sib
T = 1 →1	a	2	0
2	b	0	3
3	c	0	4
4	d	0	0
•			
•			
•			
N			

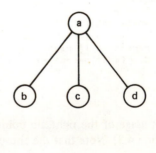

As in the list implementation of trees, a list of empty node_cells in the table is required.

4.2 Binary Trees

A very special form of trees which has more uses in practice and is easier to implement is the *binary tree*. A binary tree is a tree where each node can have a maximum of two subtrees.

Definition. A binary tree is a collection of nodes with the following hierarchical structure:

    either (a) it is an empty tree structure
    or   (b) it has a root node with
                (i)   a left subtree (binary)
                (ii) a right subtree (binary)
                (iii) no subtrees
                (iv) a left and right subtree (binary)

For example

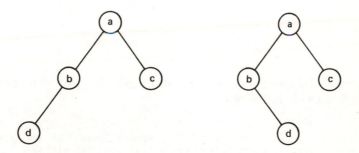

A subtree is either the left or the right subtree of its parent node and therefore the above two trees are not equivalent. Binary trees are a very important sub-set of general trees. Firstly, this is because many applications of trees use binary trees and secondly, binary trees can be used to implement general trees. A classic example of using binary trees is to represent the structure of arithmetic expressions. Figure 4.4 shows the binary tree representation of the arithmetic expression

    $J * 2 + (20 - K/3) / 13$

The simplicity of binary tree structures implies that their set of operations can be somewhat simplified. The operations initialise, empty, root, retrieve,

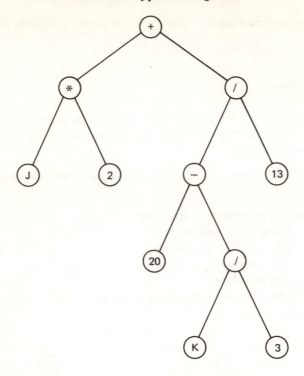

Figure 4.4   A binary tree expression.

parent, null_node, delete and assign are unchanged. But, the other operations can be replaced with the following:

```
function cons_tree (E : elemtype; L, R : binary_tree) : binary_tree;
   (* A binary tree is constructed with a root node containing E
       and L and R as its children. *)
function left (N : node) : node;
function right (T : node) : node;
```

These two functions return the left and right nodes respectively of a node N. Using these operations the tree-traversal algorithms are further simplified.

```
procedure in_order (N : node);
   begin
     if not null_node(N) then
         begin
           in_order(left(N));
           (* process the node N *)
           in_order(right(N))
         end
   end;  (* of pre-order *)
```

The other two traversal algorithms are very similar. Note that the pre_order, post_order and in_order traversals of expression trees correspond to the pre-fix, post-fix (or Reverse Polish) and in-fix (without brackets) notations used for expressions.

Implementation of Binary Trees

Using arrays, a similar structure to that used for general trees is applicable. For unbounded binary trees (in size) the pointer implementation is preferred.

```
type node = ↑node_cell;
     node_cell = record
                       element : elemtype;
                       left, right : node
                   end;
     binary_tree = node;
```

Example 1: Binary trees of expressions
Given an expression in Reverse Polish, we wish to construct a binary tree corresponding to that expression.
Solution. This process is similar to the evaluation of Reverse Polish expressions with the difference that instead of evaluating each operator, we construct its tree representation instead. For instance:

10 20 + will be transformed to

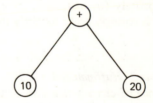

In a similar way to the case study of chapter 3, we assume that an expression in Reverse Polish notation is stored as a list of symbols. Using the same declaration of stack and list data types, the algorithm can be expressed as:

```
type kind = (no, id, op);
     (* symbol is the type corresponding to elemtype
        and refers to a symbol which is either an
        operator, a number or an identifier. *)
     symbol = record
                  case sym_kind : kind of
                  no : (value : integer);
                  id : (identifier : char);
                  op : (operator : char)
              end;
var S : stack;                        (* stack of binary trees *)
```

```
      emp, T, T1, T2 : binary_tree;    (* binary trees of symbols *)
      exp : list;                      (* the R.P. expression *)
      p : position;
      sym : symbol;

begin
  initialise_stk(S);
  initialise_tree(emp);  (* an empty tree *)
  (* An expression is read, transformed to R.P. and
     is stored in the list exp *)
  p := first(exp);
  while p <> last(exp) do
    begin
      sym := retrieve_list(exp,p);
      case sym.sym_kind of
          no, id : push(S, cons_tree(sym, emp, emp));
          op     : begin
                     T1 := top(S);  pop(S);
                     T2 := top(S);  pop(S);
                     push(S, cons_tree(sym, T2, T1))
                   end
          end;  (* of case *)
      p := next(p,exp)
    end;    (* of while *)
  T := top(S);
(* T is the binary_tree representation of the expression *)
end;
```

In the above program segment, operations initialise and retreive are post-fixed appropriately (by _lst, _stk, _tree) to remove any ambiguity of meaning. The reader is strongly urged to verify the mechanics of this algorithm.

Example 2: Huffman codes

Sequence of characters are to be transmitted across a communication link. The link is capable of transmitting binary digits 0 and 1. It is required to code characters before transmitting them so that the total length of messages transmitted is minimal. This is to ensure maximum throughput. However, it must be relatively easy to decode these codes at the other end of the communication link. The probability of occurrence of each character is known independently of other characters in advance. For example, the characters are a, b, c, d and e with the following probabilities:

character	probability
a	0.3
b	0.05
c	0.3
d	0.25
e	0.1

If fixed-length codes are used, at least 3 bits will be required for each code. For instance, a possible set of codes will be

character	*code*
a	000
b	001
c	010
d	011
e	111

The average code length is therefore 3 bits long. If variable-length codes are used, it must be ensured that no code is a pre-fix of another code. This is to enable us to decode messages without ambiguity. If, for example, the code for a is '01', the code for b is '010' and the code for c is '101', the message '01010101' can be interpreted either as 'aaaa' or 'bca'. If no code is a pre-fix of another, this ambiguity will not arise. One such coding is

character	*code*
a	111
b	01
c	10
d	110
e	00

The average code is now

$$\sum_{\substack{\text{all character} \\ \text{codes}}} \text{length} \times \text{probability} = 3\,(0.3 + 0.25) + 2\,(0.05 + 0.3 + 0.1)$$

$$= 1.65 + 0.9 = 2.55$$

Is this the optimal average code length? Before answering this question, let us define a special type of node in a binary tree.

External nodes. In a binary tree we replace every empty subtree with a special node. Such special nodes are called *external* nodes. Figure 4.5 shows the external nodes as rectangular boxes. Other nodes are shown as circles.

A *path* from the root to an external node is the sequence of nodes $n_1, n_2, \ldots, n_k$ where n_{i+1} is a child of n_i, n_1 is the root and n_k is an external node. The length of this path is $k-1$. For instance, (X, Y, Z, 1) is the path from X to external node 1. The length of this path is 3.

Let us assume that the external nodes 1, 4, 3, 2 and 5 in figure 4.5 correspond to characters a, b, c, d and e respectively. We can construct codes for these external nodes by a simple procedure. We mark a link from a node to its left subtree as '1' and a link to its right subtree as '0'. A code for an external node simply corresponds to the sequence of marks on a path from the root to that node. Figure 4.6 shows an example of such a marking scheme.

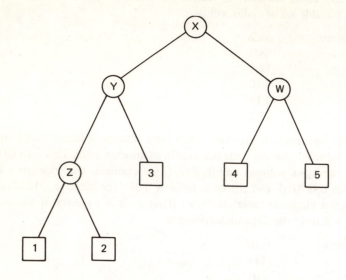

Figure 4.5   External nodes of a binary tree.

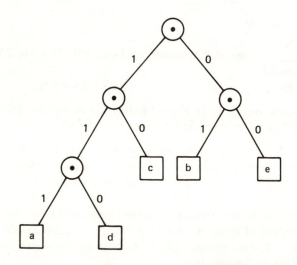

Figure 4.6   Marking scheme for binary trees.

The code of a is therefore '111'. The code of b is '01' and so on. Indeed, this tree produces the codes we gave before. Since each code corresponds to a path from the root to an external node, no code can be a pre-fix of another. To ensure that these codes are optimal, let us pose the following problem:

Given m weights $w_1, w_2, \ldots, w_m$, construct a tree with these m weights as its external nodes, such that the weighted external path length of the tree is minimal; that is

$$\sum_{\substack{i \text{ is an} \\ \text{external node}}} w_i \times L_i \text{ is minimal}$$

where L_i is the path length of external node i.

Huffman Algorithm

The minimal tree can be constructed in the following way:

(1) Choose the two smallest weights, say w_i and w_j.
(2) Replace w_i and w_j by $w_i + w_j$ and construct

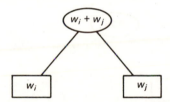

(3) Repeat steps (1) and (2) until the sequence is reduced to one weight (sum of all the w values).

Let us apply this algorithm to the probabilities of our five characters. The two smallest weights are 0.05 and 0.1. Thus we replace them by $0.05 + 0.1 = 0.15$. Then the two smallest are 0.15 and 0.25 which are replaced by 0.40. The next two smallest weights are 0.3 and 0.3 which are replaced by 0.6, and finally 0.4 and 0.6 are replaced by 1.00.

(i)

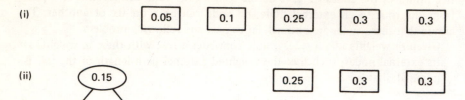

(ii)

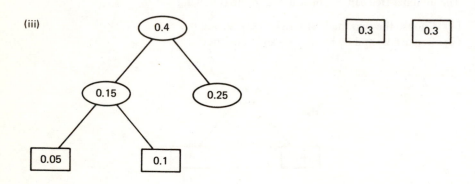

(iii)

(iv)

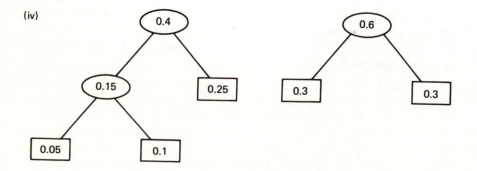

(v)

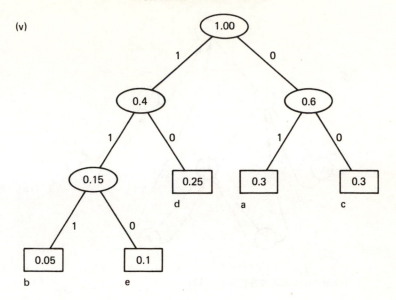

The optimum codes can therefore be derived as

character	code
a	01
b	111
c	00
d	10
e	110

The average code length = 3 (0.05 + 0.1) + 2 (0.25 + 0.3 + 0.3)
                                      = 2.05.

For implementation of Huffman algorithm see exercises **4.15** and **4.16**.

4.3   Threaded Binary Trees

In the implementation of tree traversals, recursive procedures were used. Recall from the previous chapter that recursive procedures use an implicit stack. The size of the stack needed is proportional to the maximum path length of the tree. For large trees, this extra storage may be excessive. A simple way to traverse trees without using a stack is by using *threads*. Threads are the empty subtrees (*nil* pointers) which are used for specific purposes. One way to organise a threaded tree is to use a left nil pointer to point to the previous node and a right nil pointer to point to the next node in the *in-order* traversal of the tree. Figure 4.7 shows an example of a threaded tree. The dotted lines are threads.

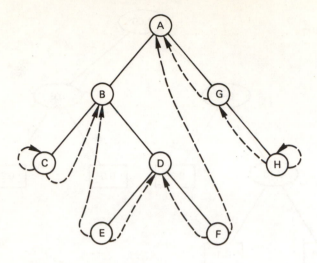

in-order traversal: C B E D F A G H

Figure 4.7 A threaded tree.

The right pointer of a node with no successor node is made to point to itself. Similarly, the left pointer of a node with no predecessor node is made to point to itself. To implement these trees in each node we should be able to distinguish between proper right and left subtrees and threads. To do this we include two flags in each node.

```
type node = ↑node_cell
     node_cell = record
                      element : elemtype;
                      right, left : node;
                       R_th, L_th : boolean;
                  end;
     binary_tree = node;
```

If R_th is true, then the right pointer is a thread, otherwise it is a proper right subtree. Similarly if L_th is true, the left pointer is a thread. Based on this data structure, we can write a function successor which returns the successor of a node (or the same node if there is no successor).

```
function successor (N : node) : node;
  var S : node;
  begin
    if N↑.R_th then
        successor := N↑.right
    else begin
```

```
        S := N↑.right;
        while not (S↑.L_th) do S := S↑.left;
        successor := S
      end
end;
```

The algorithm works by the simple observation that if N has a right thread, then its successor is found by following that thread. If not, the successor of N must be the 'furthest node to the left' of its right subtree (node N', say). This follows from the definition of in-order traversal (see figure 4.8).

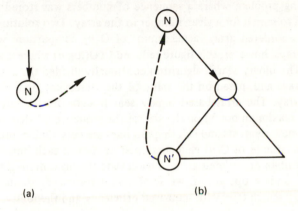

(a)   (b)

Figure 4.8   The successor of the node.

The 'furthest node to the left', N', can be found by following the left pointers until a left thread is found. The **while** loop of the algorithm performs this task. Now, the in-order traversal can be simply achieved using a loop.

```
procedure in_order (T : binary_tree);
  var s : node;
  begin
    if not empty(T) then
      begin
        s := T;
        (* to get to the first node in the sequence :- *)
        while not (s↑.L_th) do s := s↑.left;
        (* process node s *)
        while s↑.right <> s do
            begin
              s := successor(s);
              (* process node s *)
            end
      end
  end;
```

The function parent can now be written without an extra pointer in each node (see exercise **4.6**).

4.4 Binary Search Trees

In section 4.2 we discussed two applications using binary trees. A third application of binary trees is for implementing simple sets. In chapter 1 we posed a simple searching problem where a sequence of numbers was stored in an array and we needed to search for a given number in the array. Two solutions were discussed. In an unordered array, a maximum of $O(n)$ comparisons was needed. For ordered arrays, however, this figure reduced to $O(\log n)$ where n was the size of the array. The binary search algorithm continually divided the array into two smaller sub-arrays and, based on the value of the search key, it would search a half-size sub-array. The array-based binary search algorithm is only applicable when there is a maximum bound on the size of the sequence. In that implementation, inserting new elements and deleting old ones are very time-consuming. This is because, a maximum of $O(n)$ elements must be moved each time in the array in order to maintain an ordered array. To alleviate the maximum-size constraint and expensive update operations, we shall give a somewhat similar structure, namely, a binary search tree, with improved efficiency and flexibility.

In a binary search tree the nodes contain the numbers of the sequence. The nodes of the binary search tree are ordered in the following sense:

> For each non-empty subtree T, all the nodes in the right subtree of T have numbers *greater* than the root of T and all the nodes in the left subtree of T have numbers *smaller* than the root of T.

This ordering implies that at the root of tree, the sequence of numbers is partitioned into two groups: those to the left and those to the right of the root. These correspond to the left and right sub-arrays in the binary search algorithm. Figure 4.9 shows an example of a binary search tree with 9 nodes.

Also, the above ordering implies that an in-order traversal of a binary search tree would yield all the numbers in ascending order of magnitude. Assuming the following declarations, we shall now give the search, insert and delete operations.

The recursive structure of the search algorithm follows closely from the definition of binary trees. A binary tree is either empty or it has a root and two subtrees. Thus, if the tree T is empty, search for x fails. Otherwise, if x is less than the root value, the left subtree of T should be searched; if x is greater than the root value, x should be searched for in the right subtree of T; and finally if x is equal to the root value, the search succeeds. In figure 4.9 the heavy lines show the path from the root to node 25 when searching for that number.

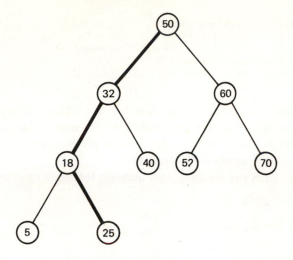

Figure 4.9   A binary search tree.

```
type node = ↑node_cell;
     node_cell = record
                     element : integer;
                     left, right : node
                  end;
     bs_tree = node;
     (* bs_tree is a pointer to the root
        node of a binary search tree *)

function search (x : integer; T : bs_tree) : boolean;
     (* if x is in the tree, a true value will be returned,
        else a false value will be returned *)
  begin
    if empty(T) then
        search := false
    else if T↑.element = x then
                search := true
         else if x < T↑.element then
                search := search(T↑.left)
              else       (* x > T↑.element *)
                search := search(T↑.right)
  end;   (* of search *)

procedure insert (x : integer; var T : bs_tree);
  begin
  if empty(T) then
        begin
          (* create a tree with one node at the root *)
          new(T);
          T↑.element := x; T↑.left := nil; T↑.right := nil
        end
    else if T↑.element = x then
                error     (* the element x is in the tree already *)
```

```
    else if x < T↑.element then
          insert(x, T↑.left)
      else                        (* x > T↑.element *)
          insert(x, T↑.right)
end;  (* of insert *)
```

To insert a new node into a binary search tree, a similar strategy can be used. Recursively, the new element is compared with the root of the tree and the left or right subtree is chosen accordingly. If an empty tree is encountered, a new tree with one node alone should be created.

The binary search tree resulting from inserting the sequence of numbers

20 30 18 2 80 22

into an empty tree is therefore:

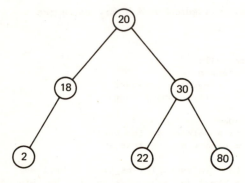

The delete operation is slightly tricky. To delete a node, first it must be located. So, the framework of the delete operation is that of the search operation. Only when the element is found can action be taken to delete it.

```
procedure delete (x : integer; var T : bs_tree);
  (* delete x from tree T *)
  begin
    if empty(T) then
          error
    else if x < T↑.element then delete(x, T↑.left)
        else if x > T↑.element then delete(x, T↑.right)
            else          (* x is located in T↑.element *)
                delete_root(T) (* delete the root node of T *)
  end; (* of delete *)
```

The delete_root procedure should delete the root of its variable parameter T. Suppose that the root of the tree T is x and it has subtrees T_L and T_R.

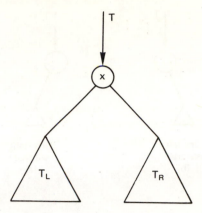

Let us use the notation $x > T$ ($x < T$) to mean that x is greater (less) than all the elements of tree T. One way to delete the root and maintain the structure of the tree is to replace it with one element of T_L (y say). Since $x > T_L$ and $x < T_R$, to preserve the invariance of the binary search tree, y must be the maximum element of tree T_L.

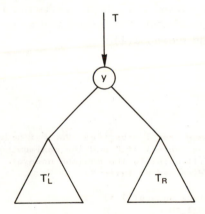

T_L' is T_L with node y removed (see procedure exchange_max below). Now, $y > T_L'$ and $T_R > y$, since y was in T_L before. So to complete the operation, we should seek the maximum element of the tree T_L. The definition of the binary search tree implies that the maximum element must be the 'furthest node to the right' of the tree, which is found by following all the right subtrees from the root until a tree with no right subtree is found. To complete the delete_root operation, we observe that there are four cases to consider:

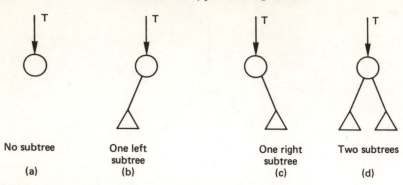

No subtree	One left subtree	One right subtree	Two subtrees
(a)	(b)	(c)	(d)

The solutions will be:

T : = nil	T := T↑.left	T := T↑.right	Follow the above process
(a)	(b)	(c)	(d)

The first three cases are trivial. For case (d), we describe a procedure exchange_ max to find the maximum element of the left subtree and to remove it after it has been copied into the root node (T).

```
procedure exchange_max (var T : bs_tree; var P : node);
    (* It exchanges the content of P and the maximum node
       of T; then P will point to the maximum node of T
       which will be disposed of later *)
    begin
      (* find the maximum node of T *)
      if T↑.right <> nil then
          exchange_max(T↑.right, P)
      else   (* the maximum node is found, i.e. node T *)
          begin
            (* Exchange contents of T and P *)
            P↑.element := T↑.element,
            T := T↑.left;        (* see diagram below *)
            P := T
          end
    end;  (* of exchange_max *)
```

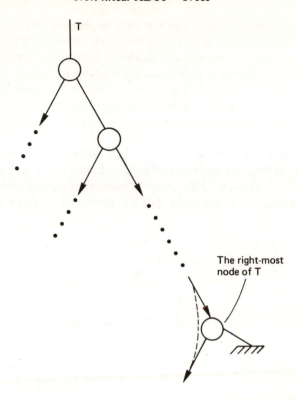

The right-most node of T

The dotted line shows the necessary modification to the tree. The complete algorithm is given below.

```
procedure delete_root (var T : bs_tree);
   var temp : bs_tree;
   begin
      temp := T;
      if T↑.right = nil then T := T↑.left
      else if T↑.left = nil then T := T↑.right
           else exchange_max(T↑.left, T);
      dispose(T)
   end;  (* of delete_root *)
```

Indeed, an alternative algorithm can be developed which would replace the node to be deleted with the minimum node of its right subtree (see exercise **4.10**). For other methods of applying the delete operation see exercises **4.11**, **4.12**, **4.13** and **4.14**.

Time Requirement of Binary Search Tree Algorithms

The three algorithms above all start at the root and walk along a path to a particular node. The worst case of the search algorithm occurs when the search element is in a leaf node or it does not exist in the tree. The insert and delete algorithms always traverse a path to a leaf node. The insert algorithm adds a new node as a leaf of the tree and therefore traverses a complete path. The delete operation traverses a path from the root node to node T to locate an element, and then it traverses a path from T to the leaf node 'furthest to the right' of the left subtree of T (in the worst case). Therefore, to analyse the time requirement of these algorithms, we need to investigate the lengths of paths to leaves.

First, let us define a few useful terms. The *height* of a tree is the maximum length of a path from the root to a leaf node. A binary tree is said to be *perfectly balanced* if, for each node of the tree, the number of nodes to the left and right of that node differ by at most one. Some examples of perfectly balanced trees are

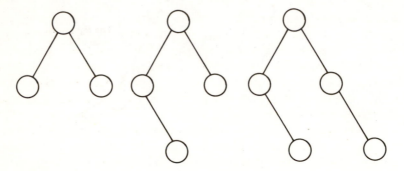

Let H(n) be the height of a perfectly balanced tree with n nodes. The $n - 1$ nodes of the subtrees should be distributed evenly in the left and right subtrees. In the worst case, one subtree has one more node than the other subtree, and we take the subtree with the greater number of nodes in it. Therefore, the heavier subtree will have $n/2$ nodes (if y is the number of nodes in the heavier subtree, $y + y - 1 = n - 1$ and therefore $y = n/2$). Thus

$$H(n) = 1 + H\left(\frac{n}{2}\right)$$

where 1 represents the path from the root node to its child. This equation yields H(n) = k + H($n/2^k$) and, since H(1) = 0, H(n) = $\log_2 n$ + 0. Therefore H(n) = O($\log_2 n$).

As with the binary search algorithm on ordered arrays, a perfectly balanced binary search tree gives a logarithmic time requirement for the search, insert and delete operations. However, not all trees have height proportional to log n. For instance, the following tree has height O(n):

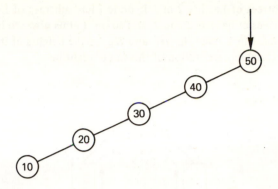

This tree would be generated by applying the insert algorithm to the sequence '50 40 30 20 10'. This, in fact, represents the maximum height of a tree with 5 nodes. So, the heights of such trees depend very much on the order in which elements are presented to the insert algorithm. It can be shown that the average height over all inputs of size n is $O(\log n)$. However, the worst case will be $O(n)$ with a *degenerate tree* as above. To ensure that all the trees generated yield logarithmic behaviour, we need to restructure our trees when inserting (or deleting) a new element. The restructuring of trees to obtain perfectly balanced trees is a very expensive process. However, with different notions of 'balancedness', we can come up with efficient restructuring techniques and, at the same time, obtain logarithmic time complexities for binary search tree operations. We shall discuss two such restructuring techniques in the next two sections.

4.5 Balanced Trees: AVL Trees

A binary search tree is *AVL balanced* if, for every node, the heights of its two subtrees differ by at most one. Examples of AVL trees are

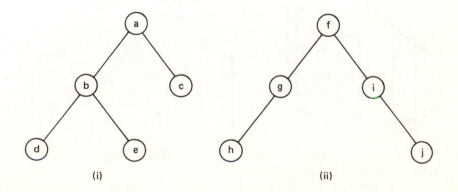

Node a has subtrees of heights 2 and 1, node i has subtrees of heights 0 and 1, node f has subtrees of heights 2 and 2, and so on. Let us now see how these trees can be generated. For a node, let H_L and H_R be the heights of its left and right subtrees. A pictorial representation of this node might be

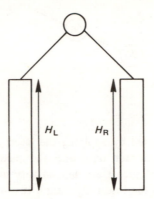

Recall that new nodes are inserted as the leaves of a tree. Let us assume that a new node has been inserted in the left subtree of the above tree and has caused its height to increase by one. Three cases may arise: if $H_L = H_R$ then the insertion of the new node does not cause any *rebalancing* of the tree; if $H_R = H_L + 1$ then the tree will be even better balanced; but if $H_L = H_R + 1$, the left subtree will have two levels more than the right subtree. This violates the AVL tree criterion and the tree must be rebalanced.

The inventors of AVL trees showed that there are primarily two distinct cases when a rebalancing is required. These two cases can be shown diagrammatically:

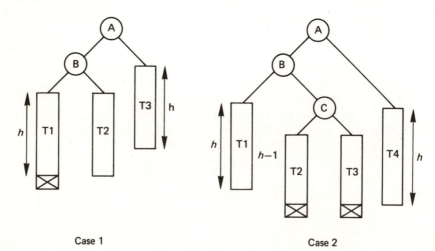

Case 1   Case 2

In case 1, the new node is inserted in the left subtree of A and has increased the height of the left subtree of B by one. The 'crossed' box indicates where the new node is added. Node B remains balanced but node A is no longer balanced. In case 2, the new node is inserted in the left subtree of A and has caused the height of the right subtree of B (either left or right subtree of C) to increase by one. Nodes C and B are still balanced but node A has become unbalanced.

The inventors of AVL trees found a very simple method of rebalancing the trees in the above cases. The solution to case 1 consists of rotating the structure clockwise and making node B the root of the new tree. The subtrees must then be exchanged as in the following diagram. In case 2, node C is moved to the root with B and A as its left and right nodes respectively. The subtrees are exchanged as follows:

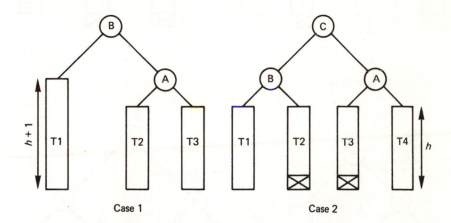

Case 1   Case 2

Note that these transformations preserve the binary search tree properties. Thus, a pre-order traversal produces keys in increasing order before and after the transformations.

There are two more cases when the tree should be rebalanced. These occur when a new key is inserted in the right subtree of the root and has increased its height by one. It happens that these two cases are symmetric cases of the above two cases. These two cases and their rebalanced trees are

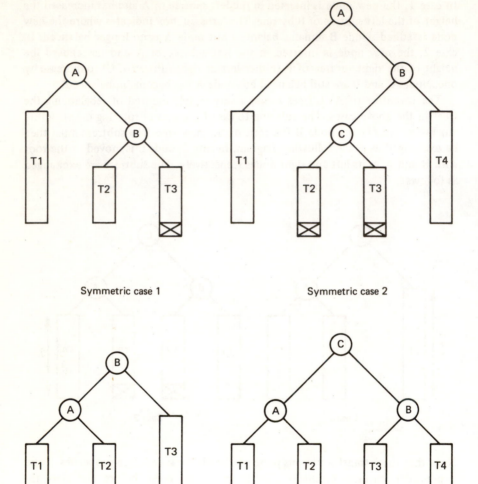

Symmetric case 1   Symmetric case 2

With these rebalancing techniques at our disposal, we can now develop the insert procedure for AVL trees. As usual, a search finds the leaf node where the new node should be inserted. Then, as we climb up towards the root following the search path (in the reverse direction), at each node a check should be made for the above cases and the tree should be rebalanced accordingly. When we use a recursive procedure for the search operation, as recursion winds up we can do our checks. Thus a layout for the insert operation is like this:

```
procedure AVL_insert (x : integer; var T : bs_tree;
                                    var increased : boolean);
   (* insert x in tree T. If the height of the tree is increased,
```

```
       the variable 'increased' will return a true value *)
begin
  if empty(T) then
    (* create a new node for x and set increased to be true *)
  else begin
        if x = T↑.element then error
        else if x > T↑.element then
              begin
                AVL_insert(x, T↑.right, increased)
                if increased then
                    check_and_rebalance
              end
            else (* x < T↑.element *)
              begin
                AVL_insert(x, T↑.left, increased)
                if increased then
                   check_and_rebalance
              end
      end
end;
```

The crucial part of the algorithm is the check_and_rebalance procedure. Given the root of a tree, it is possible to traverse the left and right subtrees of each node to find their heights and therefore check if the tree is unbalanced. However, this method is very time-consuming. A better solution is to maintain a flag in each node to indicate if

- subtrees have *equal* heights
- the left subtree is *higher* than the right subtree
- the left subtree is *shorter* than the right subtree

That is, flag can take values of equal, left_higher or right_higher. For instance

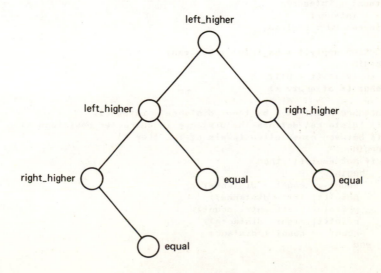

Now, an examination of cases 1 and 2 reveals that if x is the flag of node A and y is the flag of node B after inserting the new node, then the following cases may occur:

(i)   x = right_higher or equal . . . no rebalancing
(ii) x = left_higher and y = left_higher . . . case 1
(iii) x = left_higher and y = right_higher . . . case 2

A similar result holds for the symmetric cases:

(i)   x = left_higher or x = equal . . . no rebalancing
(ii) x = right_higher and y = right_higher . . . symmetric case 1
(iii) x = right_higher and y = left_higher . . . symmetric case 2

The AVL_insert algorithm can now be completed by maintaining a flag for each node. The following listing gives a program for reading a sequence of numbers terminated by a zero, inserting them into an AVL tree and then printing the resulting tree. Figure 4.10 then shows the input to and output from the program (note that the lines need to be drawn by hand).

```
program AVL_tree (input, output);
  type
    flagtype = (left_higher, right_higher, equal);
    nodeptr = ↑node;
    node = record
              element : integer;
              left, right : nodeptr;
              flag : flagtype
            end;
    bs_tree = nodeptr;
  var
    t : bs_tree;
    count : integer;
    x : integer;
    increased : boolean;

  function empty(t : bs_tree) : boolean;
    begin
      empty := (t = nil)
    end; (* of empty *)

  procedure print (t : bs_tree; distance : integer);
    (* 'distance' defines the distance in character positions *)
    (* between consecutive levels of the tree *)
    begin
      if not empty(t) then
        begin
          count := count - distance;
          print(t↑.left, distance);
          writeln(t↑.element : count);
          print(t↑.right, distance);
          count := count + distance
        end
```

```
    end; (* of print *)

procedure avl_insert (x : integer; var t : bs_tree;
                                    var increased : boolean);
    var
      R_L, R_R : nodeptr;
    begin
      if empty(t) then
        begin
          new(t);
          t↑.element := x;
          t↑.left := nil;
          t↑.right := nil;
          t↑.flag := equal;
          increased := true
        end
      else if x < t↑.element then
        begin
          avl_insert(x,t↑.left, increased);
          if increased then
            begin
              case t↑.flag of
                right_higher: begin
                                t↑.flag := equal;
                                increased := false
                              end;
                equal        : t↑.flag := left_higher;
          left_higher :
            begin
              increased := false;
              (* The height of the re-balanced
                 tree will not have increased *)
              if t↑.left↑.flag = left_higher then
                begin (* CASE 1 *)
                  R_L := t↑.left;
                  (* keep the root's left child *)
                  R_L↑.flag := equal;
                  t↑.flag := equal;
                  t↑.left := R_L↑.right;
                  R_L↑.right := t;
                  t := R_L
                end
              else if t↑.left↑.flag = right_higher then
                begin (* CASE 2 *)
                  R_L := t↑.left;
                  t↑.left := R_L↑.right↑.right;
                  R_L↑.right↑.right := t;
                  t := R_L↑.right;
                  R_L↑.right := t↑.left;
                  t↑.left := R_L;
                  if t↑.flag = left_higher then
                    begin
                      t↑.right↑.flag := right_higher;
                      t↑.left↑.flag := equal
                    end
                  else if t↑.flag = right_higher then
                    begin
```

```
                          t↑.right↑.flag := equal;
                          t↑.left↑.flag := left_higher
                      end
                  else
                      begin
                          t↑.right↑.flag := equal;
                          t↑.left↑.flag := equal
                      end;
                  t↑.flag := equal
              end
          else (* t↑.left↑.flag = equal *)
              begin
          (* Nothing to be done *)
              end
          end  (* of left_higher case *)
        end  (* of case statement *)
      end
    end
else if x > t↑.element then
  begin
      avl_insert(x,t↑.right, increased);
      if increased then
        begin
          case t↑.flag of
          left_higher : begin
                            t↑.flag := equal;
                            increased := false
                        end;
          equal       : t↑.flag := right_higher;
          right_higher:
          begin
            increased := false;
            if t↑.right↑.flag = right_higher then
              begin  (* SYMMTERIC CASE 1 *)
                R_R := t↑.right;
                (* keep the root's right child *)
                t↑.flag := equal;
                R_R↑.flag := equal;
                t↑.right := R_R↑.left;
                R_R↑.left := t;
                t := R_R
              end
            else if t↑.right↑.flag = left_higher then
              begin  (* SYMMETRIC CASE 2 *)
                R_R := t↑.right;
                t↑.right := R_R↑.left↑.left;
                R_R↑.left↑.left := t;
                t := R_R↑.left;
                R_R↑.left↑.left := t↑.right;
                t↑.right := R_R;
                if t↑.flag = right_higher then
                  begin
                    t↑.left↑.flag := left_higher;
                    t↑.right↑.flag := equal
                  end
                else if t↑.flag = left_higher then
                  begin
```

```
                              t↑.left↑.flag := equal;
                              t↑.right↑.flag := right_higher
                           end
                        else
                           begin
                              t↑.left↑.flag := equal;
                              t↑.right↑.flag := equal
                           end;
                         t↑.flag := equal
                     end
                  else  (* t↑.right↑.flag = equal *)
                     begin
               (* Nothing to be done *)
                     end
                  end
               end
            end
         end
   else  (* x = t↑.element *)
      increased := false
   end; (* of avl_insert *)


(* THE MAIN PROGRAM *)
begin
   count := 70;          (* The maximum length of a line *)
   (* Read a sequence of numbers terminated by a zero and *)
   (* insert them in tree t *)

   t := nil;             (* t is an empty tree *)
   writeln('Type a sequence of non-zero integers to be terminated by a zero'
   read(x);
   while x <> 0 do
      begin
         avl_insert(x,t,increased);
         read(x)
      end;
   writeln;
   writeln;
   writeln('Now, the AVL tree is displayed :-');
   print(t,7)
end.
```

Figure 4.11 shows the actions of the algorithm. Numbers 1 and 2 are inserted without any rebalancing. When node 3 is added ((i), below), node 2 is still balanced but node 1 is not. Nodes 1 and 2 correspond to nodes A and B respectively in the symmetric case 1, and thus a simple rotation moves node 2 to the root with nodes 1 and 3 as its left and right subtrees ((ii) below). When node 5 is inserted, another rotation at node 3 is necessary. When node 6 is added, node 2 will be unbalanced ((iii) below). This is the symmetric case 2 where nodes 2, 4 and 3 correspond to nodes A, B and C respectively. A rotation of type 2 therefore preserves the balance of the tree and so on.

The mathematical timing analysis of AVL trees is very difficult. However, empirical results show that the height of these trees is very close to log *n*. How-

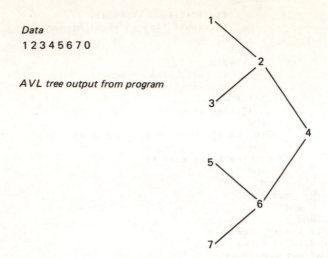

Data

1 2 3 4 5 6 7 0

AVL tree output from program

Figure 4.10 Sample data and its output from the AVL program.

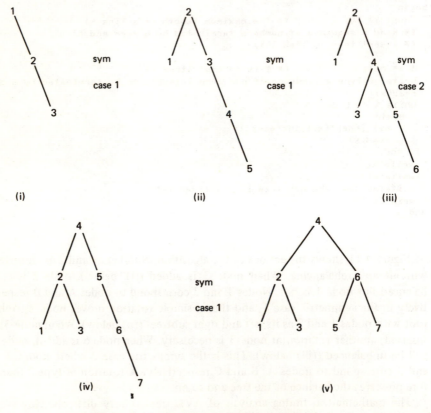

Figure 4.11 Sequence of rebalancings.

ever, in the insert algorithm, extra work is needed to rebalance the tree. Experimental results show that, on average, every two insertions require one rebalancing. Therefore, for trees where the frequency of the searching operation is much higher than the frequencies of the insert and delete operations, the AVL tree is preferred. This argument is equally valid for other balanced trees (see next section). The delete operation on AVL trees is very similar to the insert operation.

4.6 Balanced Trees: B-trees

In this section we introduce another balanced tree implementation technique. We shall begin by examining the possibility of organising tree structures on backing storage devices such as magnetic discs. On these devices, the unit of data transfer is one *block* of data. A block is typically 512 bytes of memory. The nature of magnetic discs is that the access time of a block of data is, among other factors, dependent on the position of the block on the disc. Reading a block of memory firstly requires time to move the reading head of the disc to the exact location. The average access time is usually in the order of 50 milliseconds. If we organise a binary search tree on a disc memory, accessing a child of a node may require a block access on the disc (in the worst case). For a balanced tree, it may be necessary to access (log n) blocks of the disc (on average). In practice, however, the elements of a tree are fairly small (compared with the size of a block) and it would be advantageous to cluster together many related nodes of a tree into one block and consequently to reduce the total height of a tree. One such clustering technique is used in B-trees. We shall discuss B-trees in this general context. Later on, specialised forms of B-trees will be introduced for use in the main memory.

B-trees

A B-tree of order n is defined as a tree where:

(a) Every node has a maximum of $2n$ and a minimum of n elements; except the root, which can have a minimum of one element.
(b) A node is either a leaf node with *no* children or an internal node with m elements and $m+1$ children ($n \leqslant m \leqslant 2n$). If $e_1, e_2, \ldots, e_m$ denote these elements, $P_0, P_n \ldots, P_m$ denote the children, then
   P_0 contains all elements less than e_1
   . . . .
   P_i contains all elements less than e_{i+1} and greater than e_i
   . . . .
   P_m contains all elements greater than e_m

(c) All leaves are at the same level.

A typical node of a B-tree is:

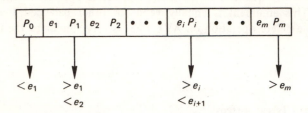

Figure 4.12 shows an example of a B-tree of order 2. A node can have between 2 and 4 elements. The root can have only one element. Note that all the leaves are at the same level.

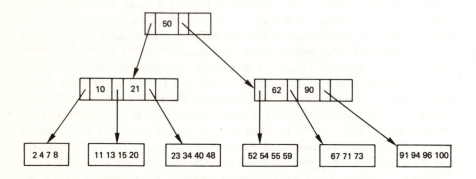

Figure 4.12   Example of a B-tree of order 2.

The search algorithm can easily be derived, based on the structure of B-trees. An outline of the search algorithm can be given as

```
function search (x : elemtype; T : B_tree) : boolean;
   begin
     if T is empty then search := false
     else
        begin
          (* Access the root node of T.  Let us
             assume that this node has elements
             e₁, e₂, ..., eₘ and children P₀, P₁, ..., Pₘ *)

          if x < e₁ then search := search(x, P₀)
          else if x > eₘ then search := search(x, Pₘ)
```

```
      else    (* in pseudo-Pascal *)
          if x = e_i for some 1 ≤ i ≤ m then
                    search := true
          else if e_i < x < e_{i+1} for some 1 ≤ i ≤ m then
                    search := search(x, P_i)
  end
end;   (* of search *)
```

As can be seen, a search within a node is required either to locate x or to find two consecutive elements between which x lies. If each node is very large, a binary search can be carried out within the node (the elements must be in an ordered sequence). However, this may not be necessary as the total time will be dominated by a block access on the disc.

As with AVL trees, the insertion and deletion algorithms on B-trees require rebalancing.

B-tree Insertion

Using the search algorithm, the leaf node N, where the new element should be inserted, can be found. If node N contains fewer than $2n$ elements, then the new element can be inserted. However, if node N is full, the tree needs to be re-balanced. Let us assume that node N will contain elements $e_1, e_2, \ldots, e_{2n}, e_{2n+1}$ after the new element is added. Furthermore, this sequence is ordered, that is

$$e_1 < e_2 < e_3 < \ldots < e_{2n} < e_{2n+1}.$$

Now we construct two nodes, N1 and N2, where N1 contains elements e_1, $e_2, \ldots, e_n$ and N2 contains elements $e_{n+2}, e_{n+3}, \ldots, e_{2n+1}$.

Node N

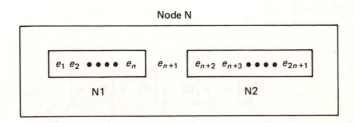

Nodes N1 and N2 have the minimum number of elements permitted. The element e_{n+1} can now move to the parent of N to be added there in its appropriate place.

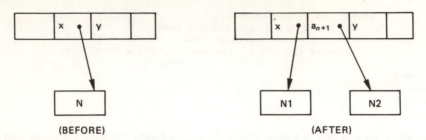

Indeed, if the parent node is also full, the same strategy should be applied to that node. Thus in the worst case, these split operations will be propagated until the root of the tree is split into two nodes and a new root is created. This way the height of the tree increases by one. It is interesting to note that B-trees grow towards the root rather than towards the leaves. To illustrate this, let us look at the steps taken in constructing a B-tree of order 2. The elements are taken from the following sequence in this order:

50   100   80   31   5

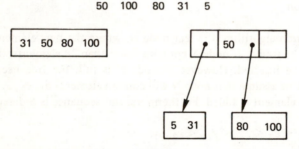

When element 5 is added, the node should be split, and a new root is created with the middle elements of the 5 elements (element 50). As another example, the B-tree of figure 4.12 will change when elements 120 and 53 are added:

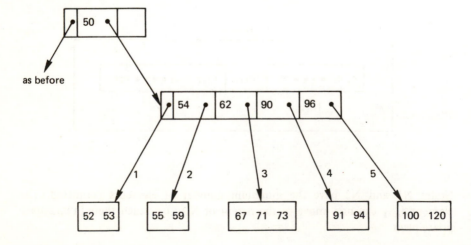

Now if elements 63 and 84 are added, the tree will be restructured as follows:

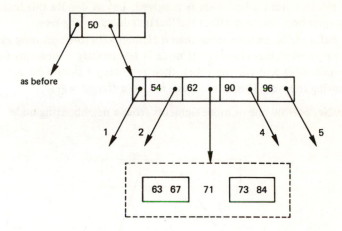

Moving 71 to the parent node is impossible since it is already full. Thus the parent node is also split:

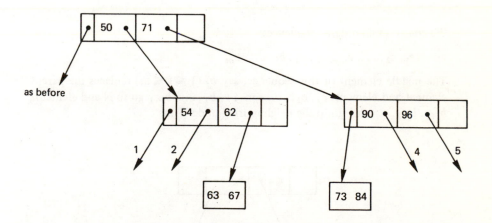

B-tree Deletion

The element to be deleted (say *e*) should be found using a search in the B-tree. Let us assume that node N contains this element. If N is a leaf node, the element can simply be removed (although, it may unbalance the tree). If not, in a similar way to binary search tree deletion algorithms, this element can be replaced with the maximum element of the immediate left child of *e*. This can be achieved by following the right-most child of the left child of *e* until a leaf node is found.

The maximum element of that node should then be replaced with *e*. After this replacement, we can remove *e* from this leaf node. In either of these cases, effectively an element from a leaf node is removed. Let us denote this leaf node by N and consider how this may affect the 'balancedness' of the tree.

If the leaf node N contains more than *n* elements (before removing *e*), there is no need for rebalancing. However, if node N has exactly *n* elements (the minimum permitted; except for the root), then removing *e* should be accompanied by rebalancing the tree. This can be done in two different ways:

1. If possible, *borrow* one or more elements from a neighbouring node.

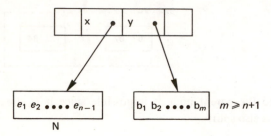

We construct an ordered sequence:

$$e_1, e_2, \ldots, e_{n-1}, y, b_1, b_2, \ldots, b_m$$

The middle element of this sequence, say b_j $(1 \leqslant j \leqslant m)$ replaces the parent element and elements $e_1, e_2, \ldots, e_{n-1}, y, b_1, \ldots, b_{j-1}$ go to N and elements $b_{j+1}, b_{j+2}, \ldots, b_m$ go to the neighbouring node. Thus

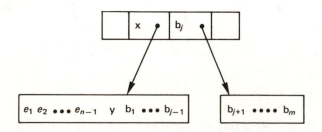

For instance, element 60 is deleted from the following tree in the following way:

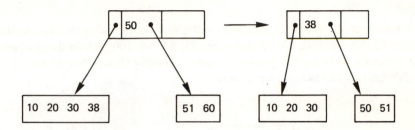

2. If the neighbouring node has exactly n nodes, naturally no element can be borrowed. In this case we merge the two nodes together to form a new node. The parent and neighbour nodes of N may look like:

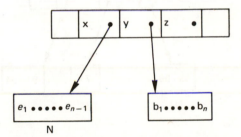

The merging is achieved by borrowing an element from the parent node. In the above configuration, element y is borrowed from the parent node. The $n-1$ elements of node N, n elements of its neighbour node and element y are put together to make a new node n with exactly $2n$ elements.

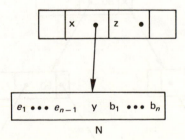

This operation implies that element y should be removed from the parent node. If the new parent node ends up with $n-1$ elements, it must be rebalanced using methods 1 and 2 outlined above. In the worst case, the merging operation of 2 above may be propagated until the root node with one element alone is merged into its only child, and thus the height of the tree is shortened by one.

Let us take the B-tree of figure 4.12 and delete the following elements in this order: 73, 52, 54, 100, 96, 67. Elements 73, 52, 54, 100 and 96 do not cause any problems. Deleting 67, however, requires merging that node with one of its neighbours (no borrowing can be done):

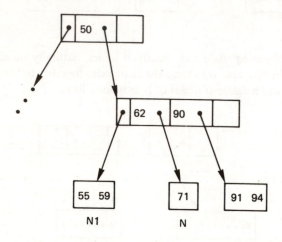

Let us merge nodes N and N1 by borrowing element 62 from their parent node. This causes the parent node to have only one element (90):

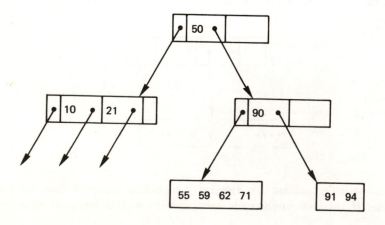

Applying the same procedure, we must merge this node and its only neighbour by borrowing element 50 from their parent node. Since 30 is the only element in the root, the new merged node will be the new root of the tree:

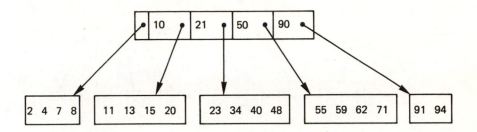

Note that a node can borrow or merge with its left or right neighbour.

Cost of B-trees

Each node of a B-tree in the worst case can be half empty. Thus the maximum storage utilisation is 50 per cent. Time complexities of the B-tree operations are directly proportional to the height of the B-tree. In the worst case, where each node is half empty (except the root), each node contains n elements and $n+1$ children. The total number of elements in a B-tree of height k is

$$T = n + n\,(n+1) + n\,(n+1)^2 + \ldots + n\,(n+1)^k$$

(There are n elements at level 0, $n(n+1)$ at level 1 etc.)
Thus

$$T = (n+1)^{k+1} - 1$$
$$T \approx n^k$$

Therefore

$$k \approx \log_n T.$$

That is, the height of a B-tree is a logarithmic function of its size T.

Variants of B-trees

The B-tree structure allows processing of elements in a random order. If, however, sequential processing is required, a simple method is to perform a slightly modified in-order traversal of the tree. This operation needs a stack of size $\log_n T$ (maximum) for the nodes.

   To remove the stack for sequential processing, a variant of B-trees known as B^+-trees may be used. In B^+-trees the actual elements are stored in the leaves. The internal nodes contain only the keys of the elements. The actual elements in the leaves can now be linked together into a linear list and in this way fast sequential access is provided to all the elements. An example of a B$^+$-tree is

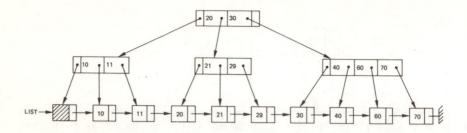

An advantage of B$^+$-trees is that, since only the keys of the elements are stored in the internal nodes, each node can store many more elements than before (since the keys are usually much smaller in size than the complete records). This in turn means that the order of tree n will be much larger. As a result, the heights of the B$^+$-trees are shorter than the corresponding B-trees, and therefore faster access times are possible.

In B$^+$-trees, it is no longer necessary to delete the internal keys, even though an element in the leaves is deleted. The key can remain and act as an index for the leaf elements. This may facilitate deletion of the elements.

A second variant of B-trees, known as B*-trees aims at improving the storage utilisation. In B*-trees, each node (except the root) must have at least $4/3n$ elements and at most $2n$ elements. That is, each node must be at least two-thirds full. The worst-case storage utilisation therefore improves to 66 per cent. This also means that the height of B*-trees will generally be shorter than the height of their B-tree counterparts with the same number of elements. However, the rebalancing of B*-trees is more complex and costly (see exercise **4.25**).

Another variant, known as compact B-trees, improves storage utilisation by as much as 90 per cent, at the cost of very inefficient insert/delete operations.

The B-tree has been primarily introduced as a data structuring technique for backing disc memories. However, it can also be used for balanced tree organisation in the main memory. When used in the main memory, it is no longer necessary to have nodes corresponding to blocks of disc (in size). Thus, a simpler form of B-trees, often known as *2-3 trees*, can be used. A 2-3 tree is a B-tree of order 1. Therefore, each node has between 1 and 2 elements and between 2 and 3 children (which explains its name).

Exercises

4.1. List the pre-order, in-order and post-order traversals of the following tree:

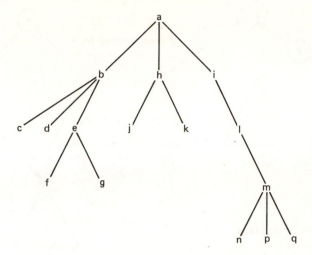

4.2. Write an in-order traversal algorithm which, given a tree expression, would list an expression with corresponding parentheses.

4.3. The cons_tree function on trees defines a series of such functions. Try to specify the operation to construct trees such that it has a fixed number of parameters. Consequently, only one function would be sufficient. (This may necessitate a change in the definition of trees.)

4.4. Give the necessary data declaration for the left_child right_sibling representation of trees using an extra flag and a thread. Complete the implementation of the tree operations.

4.5. Implement trees using binary trees.

4.6. Write a Pascal function to find the parent of a node in a threaded binary tree. What is the time requirement for this function?

4.7. Investigate insert and delete operations in a threaded binary tree.

4.8. What is the maximum time complexity of the in-order traversal algorithm in threaded binary trees? Assume that the height of the tree is h. (Derive the time requirement in terms of n, the size of the tree and h.) What is the time requirement if the tree is balanced?

4.9. Define the binary threaded tree as an abstract data type.

4.10. Write the delete procedure for binary search trees where a node is replaced with the minimum element of its right subtree.

4.11. The delete operation of binary search trees is *asymmetric*. A *symmetric* delete operation would alternate between replacing the node with the minimum element of its right subtree and the maximum element of its left subtree. Investigate the merits or otherwise of the symmetric delete operation.

4.12. The delete operation of binary search trees can be achieved by moving entire subtrees as follows:

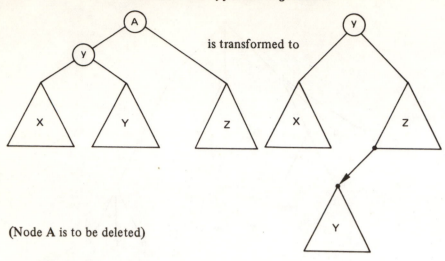

is transformed to

(Node A is to be deleted)

Rewrite the delete procedure using this technique.

4.13. What is the time complexity of your solution to exercise **4.12**?

4.14. Investigate the merits or pitfalls of the above delete algorithm in comparison with the one presented in this chapter.

4.15. Implement Huffman's algorithm using ADTs, binary trees and ordered lists of binary trees.

4.16 Find the optimal binary codes for the following set of characters with given relative frequencies. What is the average code length?

a	1
b	2
c	9
d	16
e	25
f	36
g	49
h	64
i	81
j	100

4.17 *Proof of Huffman's Algorithm*

Suppose $m \geqslant 2$ and $w_1 \leqslant w_2 \leqslant w_3 \leqslant \ldots \leqslant w_m$. Since there are a finite number of trees with m external nodes, there must be one which minimises the Huffman sum (weighted external path length). Let V be an internal node at maximum distance from the root. If w_1 and w_2 are not the weights already attached to sons of V, we can interchange them with the values already there and not increase the Huffman sum. Therefore, there is a tree which minimises the Huffman sum and contains

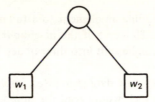

Now, it is easy to prove that the weighted path length of such a tree is minimised if and only if the tree with

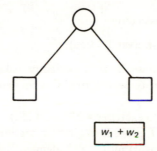

replaced by $\boxed{w_1 + w_2}$

has a minimal Huffman's sum. The same argument can be used to show that the Huffman's tree with weights $w_1 + w_2, w_3, \ldots, w_m$ is optimal.

4.18. Using pointer-representation of binary trees, complete the implementation of binary tree operations.

4.19. In order to deal with the delete operation in trees, add appropriate operations to treat the disposal of nodes properly.

4.20. Construct AVL-balanced trees from the following sequence of elements taken in this order:

> (i)   5 8 15 3 2 4 9 7 6
> (ii) 1 2   3 4 5 6 7 8 9
> (iii) 1 9   8 2 7 3 6 5 4 3 2 1

In each case, also build the corresponding binary search tree. How do they compare in their heights?

4.21. Show that a binary tree with n internal nodes has $n + 1$ external nodes. (*Hint:* first calculate the number of edges (arcs) in a tree with m nodes.)

4.22. Insert the following elements in an empty B-tree of order 2.

> 12 100   81 19   5 6 75 12 66 120
> 54   2 210 70 60 1   3 10 27   90

4.23. In the B-tree of exercise **4.22**, delete the following elements in this order:

> 1 2 3 70 75 90 100 120 210

4.24. Construct a B⁺-tree from the B-tree of exercise **4.22**.

4.25. In a B*-tree, when adding an element to a full node, a neighbouring node should be filled first. However, if the neighbouring nodes are all full, two neighbouring nodes can be split into three nodes. Write the insert algorithm for B*-trees.

4.26. Using appropriate dynamic data structures, implement 2–3 trees in Pascal.

4.27. Write an algorithm which would read the number of nodes N and N numbers, and construct a perfectly balanced tree. (Note that the binary tree need not be ordered.)

4.28. An ordered list of elements L is given (size(L) = n). Write an algorithm to construct a perfectly balanced binary search tree containing the elements of the list L.

4.29. Design and write a procedure:

> **procedure** balance(**var** T : tree);

which would reorganise T as a perfectly balanced binary search tree (assume that the binary search tree T is implemented using pointers of Pascal).

(*Hint:* use solutions to exercises **4.27** and **4.28**.)

4.30. In the AVL_INSERT procedure given in section 4.5 the two sections of code for (x < t↑.element) and (x > t↑.element) are very similar in structure. Try to parametrise this section of code so that it can be written as a procedure and then be called twice for the above cases.

4.31. Define the ADT tree as a root node and a *list* of trees. Using this specification, implement the tree operations.

Bibliographic Notes and Further Reading

An empirical study of symmetric tree deletion is given in Eppinger (1983). The results show that only for very large trees does the symmetric deletion produce better trees than random trees. The AVL trees were first introduced by Adel'son-Velskii and Landis (1962). There are some new balanced trees which compare favourably with AVL trees. Among these are k-balanced and ∈-balanced trees, Gonnet (1983).

B-trees were first introduced by Bayer and McCreight (1972). The compact B-trees and other variations are discussed in ACM (1984) and Comer (1975). The B$^+$-tree version of 2–3 trees is covered at length in Wirth (1976).

Extensive use of trees in the design of compilers for high-level languages can be found in Adel'son-Velskii and Landis (1962). Trees are also used for the design of databases. The IBM's IMS database management system is based on hierarchical structures (Date, 1985).

The problem in exercise **4.29** can be solved in O(n) time without any extra storage (Stout and Warren, 1986).

ACM (1984). *ACM-SIGMOID*, Vol. 14, No. 2.

Adel'son-Velskii, G. M. and Landis, Y. M. (1962). 'An algorithm for the representation of information', *Soviet Math.*, Vol. 3, pp. 1259–1262.

Aho, A V., Sethi, R. and Ullman, J. P. (1986) *Compilers – Principles, Techniques and Tools*, Addison-Wesley, Reading, Massachusetts.

Bayer, R. and McCreight, C. (1972). 'Organisation and maintenance of large ordered indexes', *Acta Informatica*, Vol. 1, No. 3, pp. 173–189.

Comer, D. (1979). 'The ubiquitous B-tree', *ACM Computing Surveys*, Vol. 11, No. 2, June, p. 121.

Date, C. J. (1985). *Introduction to Database Systems, Vol. 1*, 4th edition, Addison-Wesley, Reading, Massachusetts.

Eppinger, J. (1983). 'An empirical study of insertion and deletion in binary search trees', *CACM*, Vol. 26, No. 9, September, pp. 663–670.

Gonnet, G. H. (1983). 'Balancing binary trees by internal path reductions', *CACM*, Vol. 26, No. 12, December, pp. 1074–1082.

Stout, Q. F. and Warren, B. L. (1986). 'Tree rebalancing in optimal time and space', *CACM*, Vol. 29, No. 9 pp. 902–908.

Wirth, N. (1976). *Algorithms + Data Structures = Programs*, Prentice-Hall, Englewood Cliffs, New Jersey.

5 Abstract Data Type Sets—I

Sets are mathematical objects with well-studied properties and operations in modern mathematics. Sets and their operations are used extensively across a wide range of disciplines. Like abstract data types, a set is a collection of objects, usually of the same type, subject to the usual set operations such as union, intersection etc. To start with, let us illustrate how sets may be used as data types by considering a simple application.

General symbolic information can be represented using a *set of triples*. Each triple is simply an ordered triplet of identifiers. A few simple facts are represented in such a form in table 5.1. The first identifier of each triple may be treated as an object. The second and third identifiers represent an attribute and its value for a given object. The set of triples in table 5.1 represents the facts that 'John takes courses CC205 and CC206 and is 25 years old; Fred takes CC205 and his age is 30; Joe takes CC205; and, finally, Jim is an engineer and earns £12 000.

Table 5.1 A set of triples for representing general symbolic information

Object	*Attribute*	*Value*
John	course	CC206
John	course	CC205
Fred	course	CC205
Joe	course	CC205
Fred	age	30
John	age	25
Jim	is-a	engineer
Jim	salary	12 000

This set can be treated as an abstract data type 'TRIPLE_SET' with a number of operations acting on the set. Among these operations we may define

> **function** object_of (attr, val : identifier) : zet;
> (∗ Zet is defined as an ADT set of identifiers. Zet is used to distinguish it from the Pascal type set ∗)

where a set of identifiers (such as obj) is returned so that the triple (obj, attr, val) exists in the TRIPLE_SET. Note that the function 'object_of' returns a value of type *zet* which is defined to be a set of single identifiers. The ADT *zet* will have the usual mathematical operations such as union, intersection, set_difference and

142

so on. Using the above function and set operations we can create more complex sets. The set of all students taking the course CC206 can be constructed by

CC206_student := object_of ('course', 'CC206')

Naturally, CC206_student must be defined as a variable of type zet. Similarly, other sets can be built:

CC205_student := object_of ('course', 'CC205');

CC205_or_CC206 := union (CC205_student, CC206_student)
   (* This is the set of students who take CC206, CC205 or both courses *)

CC206_and_CC205 := interesection (CC205_student, CC206_student)
   (* This is the set of students who take both CC206 and CC205 *)

Age_25 := object_of ('age', '25');
   (* Age_25 is the set of all students aged 25 *)

Query := intersection (Age_25, CC206_and_CC205)
   (* Query is the set of students aged 25 who take both CC205 and CC206 *)

It is assumed that sets may be assigned to set variables (see below for the operation 'assign'), and union and intersection are functions returning the union and intersection of their two set parameters.

The simplicity and clarity of sets and their operations make them ideal data types for a variety of applications. In the next section we shall have a closer look at the operations that set data types should support.

5.1 ADT Set: Definitions

As an ADT, a set is a collection of *unique* objects of the same type (the generic type elemtype). Compared with lists and trees, sets have no *structure* in the sense that there are no 'next' and 'previous' elements for a given element. However, the elements of a set may be *ordered*. It is meaningful therefore, to state the minimum or maximum elements of a set. The operations on sets should provide for: inserting and deleting elements to and from sets; union, intersection and set difference operations; and a few functions for checking empty sets and set membership. The following gives a general list of operations for sets. More specialised operations can be added, depending on the nature of an application which uses sets or the nature of the set itself (such as the TRIPLE_SET above). In order not to confuse our ADT set with the reserved word 'set' of Pascal, we shall use 'zet' to refer to the general definition of sets.

procedure initialise (**var** S : zet);
   (* This procedure returns an empty set *)

procedure insert (x : elemtype; **var** S : zet);
 (∗ x is added to set S. Note that set elements are unique, therefore if x is already in S, no action will take place. In many algorithms in this book, we shall use the following abbreviations:

 s := { } instead of initialise(s) and
 s := {a} instead of initialise(s); insert(a,s) ∗)

procedure delete (x : elemtype; **var** S : zet);
 (∗ If element x belongs to S, it is removed from set S ∗)

function member (x : elemtype; S : zet) : boolean;
 (∗ If x is a member of S, a true value will be returned; otherwise, a false value will be returned ∗)

function equal (S1, S2 : zet) : boolean;
 (∗ A true value will be returned if S1 and S2 are equal; that is, if they contain the same elements ∗)

function empty (S : zet) : boolean;
 (∗ If S is an empty set, a true value is returned ∗)

procedure union (S1, S2 : zet; **var** S3 : zet)
 (∗ Set S3 is returned as the union of sets S1 and S2 ∗)

procedure intersection (S1, S2 : zets; **var** S3 : zet);
 (∗ Set S3 is returned as the intersection of sets S1 and S2 ∗)

procedure set_difference (S1, S2 : zet; **var** S3 : zet);
 (∗ S3 will contain elements which are in set S1 but not in set S2 ∗)

function min (S : zet) : elemtype;
function max (S : zet) : elemtype;
 (∗ These two functions return the minimum and maximum elements of a *non-empty* set S respectively ∗)

procedure assign (S1 : zet; **var** S2 : zet);
 (∗ This procedure copies set S1 to S2. Set S1 will be unaffected ∗)

function subset (S1, S2 : zet) : boolean;
 (∗ If S1 is a subset of S2, a true value is returned, otherwise a false is returned. S1 is a subset of S2 if every element of S1 is also an element (member) of S2 ∗)

procedure process_all (S : zet; **procedure** P (**var** e : elemtype));
 (∗ This procedure applies procedure P to all the elements of S. This is to implement the abstract control construct:

  for all x **in** S **do** P(x).

For instance, if procedure print(e) is defined as printing the element e, then the call process_all (S, print) would print all the elements of the set. If the elements of the set can be ordered, we may define process_all to apply p to the elements of S in that order *)

function size (S : zet) : integer;
    (* This procedure returns the number of elements in a set. Note that if a set S is empty, then size(S) = 0 *)

It is worth mentioning that the operations union, intersection and set_difference are represented as procedures rather than functions because of the restrictions which Pascal imposes on the values that a function may return. Also, the procedure assign is necessary since if, for example, an implementation of sets uses pointers, then 'S2 := S1' could have disastrous side-effects (see chapter 3 on lists). The whole structure of S1 should be copied (duplicated) for S2. The above repertoire of operations is a comprehensive set of operations for the ADT set. In later chapters we shall define somewhat more specialised set operations.

5.2 Implementing Sets using Bit-vectors

In this section we describe the simplest way of implementing sets. If all the sets are subsets of a fairly small 'universal set', then a set can be represented by a bit-vector where one bit is allocated for each element in the universal set and an element belongs to the set if its corresponding bit is set to a '1'. If an element does not belong to the set, its corresponding bit is set to a '0'. For the purpose of uniformity we assume that the universal set contains elements $1, 2, 3, \ldots,$ $N-1, N$; that is, the set of positive integers from 1 to N. N is the size of the universal set. A set is therefore represented as a bit-vector (a sequence of bits) where the ith bit is a '1' if element i belongs to the set. If $N = 10$, the following sets are represented as

Set	Bit-vector representation
S = {1, 3, 5}	<1 0 1 0 1 0 0 0 0 0>
S' = {3, 8, 10, 9}	<0 0 1 0 0 0 0 1 1 1>
F = {1, 2, 3, 4, 5, 6, 7, 8, 9, 10}	<1 1 1 1 1 1 1 1 1 1>
E = { }	<0 0 0 0 0 0 0 0 0 0>

Therefore, representation of a set requires $O(N)$ bits of storage — irrespective of its actual size. Let us assume that set S is represented by the bit-vector $<S_1,$ $S_2, \ldots, S_N>$. Now, we give the description of each set operation together with its corresponding worst-case time complexity:

initialise(s)	set all the N bits to '0'	$O(N)$ bit operation
insert(x, S)	set bit S_x to '1'	$O(1)$ bit operation
delete(x, S)	set bit S_x to '0'	$O(1)$ bit operation

member(x, S)	test bit S_x	O(1) bit operation
equal(S, S')	in the worst case, all the pairs $(S_1, S_1'), \ldots, (S_N, S_N')$ should be compared	O(N) bit operation
empty(S)	test that all the N bits are 0	O(N) bit operation
union(S, S', S'')	set $S_i'' = (S_i$ *logical or* S_i') (for all i)	O(N) bit operation
intersection(S, S'S'')	set $S_i'' = (S_i$ *logical and* S_i')	O(N) bit operation
set_difference(S, S'S'')	set $S_i'' = (S_i$ *logical and* (*logical not* S_i'))	O(N) bit operation
min(S)	search for the smallest i such that S_i is a '1'	O(N) bit operation (worst case)
max(S)	search for the largest i such that S_i is a '1'	O(N) bit operation (worst case)
assign(S, S')	copy S_i to S_i' (for all i)	O(N) bit operation
subset(S, S')	test that for all i, (S_i *logical and* (*logical not* S_i')) = '0'	O(N) bit operation
process_all(S, P)	for all i, if S_i is '1', execute P(i)	O($N \times T$) bit operation where the nominal time to execute P is O(T)
size(S)	the number of '1's in the bit-vector should be counted	O(N) bit operation

The operation 'empty' may be speeded up by maintaining a flag for each set indicating when it is empty.

The data type set as provided in the language Pascal is conventionally implemented as a bit-vector. Thus, the universal set should be specified as an enumerate type. Since most machine architectures possess instructions for bit-wise logical *and* and *or* on a pair of memory words, the operations union, intersection, set_difference, assign, empty, initialise and equal may be implemented in a time of O(N/w), where w is the size of each word (in bits). Most Pascal implementations set a maximum limit on the size of the universal set N. If larger sets are to be represented using bit-vectors, they can be implemented using a packed array of boolean values. The set operations then can be programmed in Pascal. For instance, an ADT set can be defined as

```
const N =  ; (* The size of the universal set *)
type zet = packed array [1..N] of boolean;
```

The set operations can now be easily written, for instance:

```
procedure set_difference(S1, S2 : zet; var S3 : zet);
  var i : integer;
  begin
    for i := 1 to N do
```

```
    S3[i] := S1[i] and not S2[i]
end;  (* of set_difference *)
```

The notion of bit-vectors has been used so far to represent sets of enumerate types such as sub-range of integers etc. However, the same idea can be used for implementing sets of structured objects where the objects can be somehow mapped onto a small universal set. For instance, let us take another look at the representation of expressions in the case study of chapter 3. Assuming that the type 'expression' is defined, the data about the identifiers could be represented as a *set of definition* where each definition is a pair consisting of an identifier and its corresponding expression.

```
type   expression = .... ;
       definition = record
                        id : identifier
                        value : expression
                    end;
     set_of_definition = zet;  (* set of definition *)
var    table : set_of_definition;
```

Set_of_definition is an abstract data type and operations insert and delete will apply to it. The function member may be redefined so that, if a definition belongs to the set, its entire definition is returned. Note that definitions are uniquely determined by their 'id' fields. Using this new definition of table, the operations of the program case_study of section 3.4 can be abstracted and thus simplified. When implementing the ADT set as used in this context, the choice of type 'identifier' will influence the implementation decisions immensely. If identifiers can be any character strings (starting with a letter and then followed by alphanumeric characters as in Pascal identifiers), then a more complex implementation, as described in the remainder of this chapter and the next chapter, should be used. However, if identifiers are limited to the universal set {'A', 'B , . . ., 'Y', 'Z'} then a vector of definitions can be used. It is known that the maximum number of definitions at any point in time is 26. Therefore, an **array of** definition can be used. In this way, all the simplicity of bit-vector implementations would be exploited. If the elements of a set are very large, a good strategy is to have a vector (an array) of pointers to actual elements. This will mean that, for non-existent entries, a nil pointer will be stored. In chapter 3 we used these ideas without any reference to the ADT set. The new approach, however, allows a degree of evolvability where identifiers can be strings of length more than 1 (see exercise **5.5**).

5.3   Implementing Sets using Lists

A set $\{a_1, a_2, . . ., a_n\}$ can be represented as a list $a_1, a_2, . . ., a_n$. Thus any list implementation can be used to implement a set. For unbounded sets where there is no maximum limit on the size of sets, pointer implementation of lists should be used. There are two ways in which lists can be used to represent a set. The elements of a set can be stored in an unordered list or in an ordered list. In the former case, the order of the elements in the list may be random, whereas in the latter case, elements are stored in the order of increasing magnitude of elements — that is, each element is greater than its 'previous' element and smaller than its 'next' element. Taking the following two sets of numbers, let us look at pointer-implementation of unordered and ordered lists.

$$S1 = \{1, 20, 10, 31\}$$
$$S2 = \{10, 2, 18, 19, 41, 40\}$$

Unordered Lists

The sets S1 and S2 may be represented as two linked lists (with a header cell):

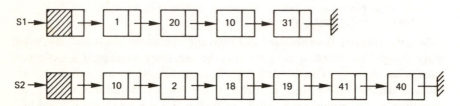

The operations initialise, empty and insert require $O(1)$ time since a new element can be added to the beginning of a list. The operations delete and member require, in the worst case, a time of $O(n)$ since a linear search is needed to find a particular element. The average-case time would be $O(n/2)$ (n is the size of a set). Operations min, max, and assign need a time of $O(n)$. Process_all needs $O(n) + M$, where M is the time needed for procedure P to process all the elements. The function size requires a complete scan of the list to count the elements, and thus needs $O(n)$ time. However, if the operation is used extensively, the size of the set can be maintained as elements are added and deleted from the set. The remaining operations equal, union, intersection, set_difference and subset need a maximum time of $O(n^2)$. This is because each operation will have to be of the general form:

```
for all x in S1 do
    (* search set S2 and process *)
```

For instance, the set_difference and intersection procedures can be written as

```
Set_difference (S1, S2, S3) :
    S3 := {};
    for all x in S1 do
```

```
   if not member(x, S2) then
       insert(x, S3);


Intersection (S1, S2, S3) :
   S3 := ();
   for all x in S1 do
       if member(x,S2) then
           insert(x,S3);
```

Using the list operations initialise, first, last and retrieve and the operation member (of sets), the loop of the intersection operation can be represented as

```
initialise(S3);
p := first(S1);
while p <> last(S1) do
  if member(retrieve(p, S1), S2) then
    insert(S3, retrieve(P, S1), first(S3));
```

The function member takes a maximum time of $O(n)$ and the loop is executed n times (assuming that the lists S1 and S2 are of nominal size n). Thus the total time is $O(n^2)$. From this analysis, it is clear that unordered lists should be used if operations requiring a time of $O(n^2)$ are not used frequently.

Ordered Lists

The above sets S1 and S2 can be represented using ordered linked lists:

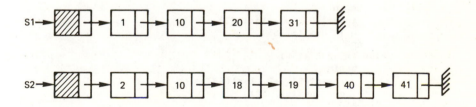

Maintaining ordered lists makes the insert operation costly. To insert a new element, first a search is needed to find the appropriate position in the ordered sequence. In the worst case, therefore, a time of $O(n)$ is needed to insert a new element into a list.

The min function can now be processed in a time of $O(1)$, since the first element of the list must be the minimum element. The max function, however, needs a complete scan of all the elements in a time of $O(n)$. If this function is used frequently, the data structures used for implementing queues can be used, which keeps track of the beginning and the end of a linked list. The main operations which are drastically affected by ordered lists are union, intersection, set_difference, subset and equal.

Let us first consider the union operation. The union of two sets represented as ordered lists can be constructed by scanning the two lists at most once. We start from the first elements of the two lists and proceed by looking at the two currently selected elements of the lists. If they are equal (z, say), z will be added to the result set and the next elements of the two lists will be selected. If the two elements are not equal (x and y, say), the smaller one (x, say) should be added to result set and its associated next element should be selected as the current element of that set, leaving the current element of the other set intact (to y). This is because the same element y could be in the other set after element x. In a single loop, such comparisons will continue until the ends of the two lists are reached. An outline of this operation would be like this:

```
Union (S1, S2, S3) :
  p := first(S1); (* p is the currently selected position in S1 *)
  q := first(S2); (* q is the currently selected position in S2 *)
  initialise(S3); (* initialise list S3 *)
  while (p <> last(S1)) and (q <> last(S2)) do
      begin
          el := retrieve(p, S1);
          e2 := retrieve(q, S2);
          if el = e2 then
              begin
                insert(S3, el, last(S3));
                p := next(p, S1);
                q := next(q, S2);
              end
          else if el < e2 then
                  begin
                    insert(S3, el, last(S3));
                    p := next(p, S1)
                  end
              else (* e2 < el *)
                  begin
                    insert(S3, e2, last(S3));
                    q := next(q, S2)
                  end
      end; (* of while *)
(* copy the rest of the elements in the larger set to S3 *)
if p = last(S1) then
  (* copy all elements at position q up to last(S2) to S3 *);
if q = last(S2) then
  (* copy all elements at position p up to last(S1) to S3 *);
```

For the two sets above, the union set S3 would be constructed in the following sequence:

Current element of set S1	Current element of set S2	Element added to the end of set S3
1	2	1
10	2	2
10	10	10
20	18	18
20	19	19
20	40	20
31	40	31
–	40	40
–	41	41

The two lists S1 and S2 are scanned only once to produce the union set S3. The time requirement is $O(m+n)$, when m and n are the sizes of S1 and S2 respectively. If $m \approx n$, then this would be $O(m+n) \approx O(2n) = O(n)$.

The other operations – equal, intersection and set_difference – are similar to union and require a time of $O(n)$ when the size of the lists is approximately equal to n. The ordered list implementation is therefore superior to unordered lists (except for the insert operation).

Finally, it should be noted that array implementation of lists can also be applied if the delete operation is not used (or used with only very low frequency).

5.4 Implementing Sets using Trees

The list implementation of sets gives a nominal linear time requirement $O(n)$. This time can be substantially reduced by using binary search trees. In chapter 4, binary search trees were introduced as a structure for representing sets of objects supporting the operations insert, delete and search. Ordinary binary search trees give an average time complexity of $O(\log n)$ and worst-case of $O(n^2)$. The more advanced balanced trees – AVL trees and B-trees – both give a guaranteed time of $O(\log n)$ for the above operations (for AVL trees, this is an experimental result). Sets S1 and S2 of section 5.3 may be represented as binary search trees:

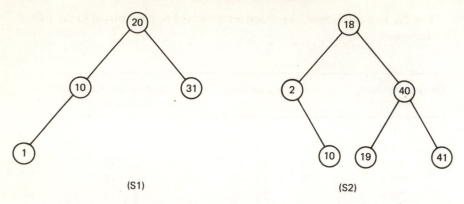

(S1)   (S2)

The operations initialise, insert, delete and empty are equivalent to tree operations. The function member is equivalent to the function search on trees. Now we shall briefly discuss how the other set operations may be realised using trees. In the discussion below, we shall assume the following declarations:

```
type bs_tree =  ; (* binary search tree as in section 4.4 *)
     zet = bs_tree;
```

min(S): The minimum element is the left-most node of the tree. Thus a simple loop will find the minimum element.

```
function min(S : zet) : elemtype;
  var N : node;
  begin
    if empty(S) then error
    else begin
          N := root(S);
          while not null_node(left(N)) do
              N := left(N);
          min = retrieve(N)
        end
  end;
```

This function traverses a path of the tree and thus requires a time of O(log *n*) on average. The max function can be written in a similar way.

Assign(S1, S2): The entire tree S1 should be duplicated. A systematic way to do this is through a traversal of the tree S1 (see procedure assign on the ADT trees in section 4.1).

Process_all(S, P): Again, any systematic traversal of the tree would visit all the elements. For instance let us use an in-order traversal:

```
procedure process_all(S : zet; procedure P(var n : node));
  var X : node;
  procedure in_order (var N : node);
    begin
      if not null_node(N) then
          begin
            X := left(N); in_order(X);
            P(N);
            X := right(N); in_order(X)
          end
    end; (* of in_order *)
  begin
    X := root(S); in_order(X)
  end; (* of process_all *)
```

Size(S): A traversal of the tree to count the nodes costs $O(n)$. Thus the size of the tree should be maintained while performing insert and delete operations, thus yielding a time of $O(1)$.

Equal(S1, S2): First it should be noted that two equal binary search trees may have different structures. For example, the following two trees are equal (that is, they represent the same set):

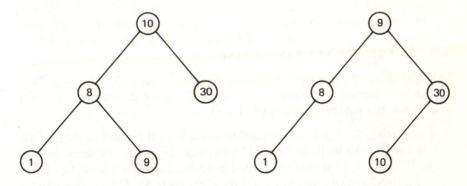

(This is only true of binary search trees, not of ordinary trees, and is because sets have no structure.)

A simple algorithm can be developed by checking the size of the two sets first and, if they are equal, then all elements of one set should be tested against the elements of the other set. Thus

```
if size(S1) <> size(S2) then equal := false
else
Check that for all x in S1, x is also a member of S2.
```

If the sets have a nominal size of n, then S1 may be scanned completely and, for each member x, a member(x, S2) operation is performed. This leads to a $O(n \log n)$ algorithm (on average).

A faster algorithm can be designed by analogy with ordered lists. If we can access all the elements of S1 and S2 in increasing order of magnitude, then a single scan of the two sequences would be sufficient. The elements of a binary search tree can be accessed in increasing order of magnitude by an in-order traversal. This follows directly from the definition of binary search trees. We cannot, however, use the recursive traversal procedure since we need to access elements of the two trees in increasing order simultaneously. Instead, we set up a non-recursive traversal of elements using an explicit stack and a function successor which would return the next element of a tree (in in-order sequence). When successor function is called for the first time for a set S, it will return the first element of the sequence (in-order).

```
equal := true;
while (more elements) and equal do
    begin
        el := successor(S1);
        e2 := successor(S2);
        if el <> e2 then equal := false
    end;
```

This approach requires a maximum time of $O(n)$. The extra cost of this algorithm is, of course, $O(2 \log n)$ storage for the stacks and their maintenance (see exercises **5.11, 5.12, 5.13, 5.14, 5.15** and **5.16**).

Union(S1, S2, S3): Copy one of the sets (S2, say) to S3 (assign(S2, S3)). Then take each element of S1 and insert it in S3 if it is not there already. For sets of size n, this requires a time of $O(n \log n)$.

Intersection(S1, S2, S3): For each element of set S1, if it is also a member of S2, insert it into the set S3 (S3 would be initially empty). In the worst case, all the elements of S1 should be inserted in S3 and therefore it needs $O(n \log n)$ to search S2 and $O(n \log n)$ to insert the elements in S3. Total time is $O(n \log n + n \log n) = O(n \log n)$.

Set_difference(S1, S2, S3): This operation is similar to intersection above, but each member of S1 should not be in S2 if it is to be added to S3.

Subset(S1, S2): All elements of S1 should be checked for membership of S2. A linearised algorithm similar to equal function operates in a time of $O(n)$.

As can be seen, tree implementation of sets gives good logarithmic behaviour for the insert, delete, member, min and max operations. This is a great improvement over lists and their linear time complexities. However, lists still remain the best structures for union, intersection, set_difference and subset operations. If necessary, however, like B^+-trees, tree data structures can be augmented (possibly with threads) so that fast sequential access becomes possible for union, intersection etc. A time of $O(\log n)$ is needed for most of the set operations.

In this chapter we have presented three general techniques for implementing sets. In the next chapter we shall study specialised sets and their novel implementations.

Exercises

5.1. Find out the value of N, the maximum size of the universal set, in your Pascal implementation of sets.

5.2. Give implementation of operations equal, union, set_difference and subset using unordered lists.

5.3. In exercise 5.2 the operations can be written using list operations alone or they can be developed directly by addressing the data structures used (ordered lists). What are the advantages and disadvantages of these two approaches?

5.4. What is the function of the following fragment of Pascal program?

```
initialise(L3); (* initialise list L3 *)
p := first(L);
while p <> last(L) do
    begin
        q := first(L2);
        while q <> last(L2) do
            begin
                if retrieve(p, L) = retrieve(q, L2) then
                    insert(L3, retrieve(p, L), first(L3));
                q := next(q,L2)
            end;
        p := next(p,L)
    end;
```

5.5. Re-implement the case study of chapter 3 using a set to store the definitions of expressions.

5.6. What is the storage requirement of exercise 5.5 in terms of size of the set of identifiers? Can it be reduced? How? (The implementation of lists and stacks need not change.)

5.7. The operation merge(S1, S2, S3) is similar to union but the sets S1 and S2 are known to be disjoint — that is, they have no common members. Write a merge operation for ordered and unordered lists. What are their time complexities?

5.8. Implement intersection, set_difference, equal and subset operations for the ordered list representation.

5.9. The union operation on ordered lists can be achieved by:

```
assign(S1, S3);
for all x in S2 do
    insert(x, S3);
```

Of course, insert(x, S3) will do nothing if x is already in S3. Write the union operation as above and compare it with the one given earlier.

5.10. Can the assign(S1, S2) be implemented using lists as a simple assignment?

> S2 := S1;

5.11. Using an explicit stack, write a successor function where

```
function successor (N : node) : node;
```

returns the next node after N in in-order traversal of a tree.

5.12. Write the function equal(S1, S2) using the successor function of exercise **5.11**.

5.13. In exercises **5.11** and **5.12**, why are stacks of minimum size O(log *n*) required?

5.14. Can the operations union, intersection and set_difference be implemented using a linearised algorithm as in the function equal when tree implementations are used? Investigate and contrast the new time requirements.

5.15. Use a linearised algorithm to implement the subset operation (as in the operation equal) for the tree representation of sets.

5.16. The linearised implementation of the function equal can be achieved by using a recursive in-order traversal of S1 and a non-recursive traversal of S2 (via successor function and an explicit stack). Write such an equal function. What are its time and space complexities?

5.17. Design data structures for threaded trees such that efficient linear access and random access to elements are possible. What are the excess costs?

Bibliographic Notes and Further Reading

Sets are used extensively in many branches of computer science. Two programming languages have been designed based on the concept of sets. The language Leap (Feldman and Rovner, 1969) uses sets of triples as discussed at the beginning of this chapter to organise a database. The language SETL (Schwartz, 1973) also uses sets as the main data type and has complex techniques for their efficient implementation.

Feldman, J. A. and Rovner, P. D. (1969). 'An algol-based associative language', *CACM*, Vol 12, No. 8, pp. 439–449.

Schwartz, T. J. (1973). *On Programming: An Interim Report on the SETL Project*, Courant Institute, New York.

6 Abstract Data Type Sets—II

In chapter 5 we looked at general implementations of the ADT set. As we noticed, not all the set operations could be implemented efficiently using a single data structure. However, many applications using sets require only a limited number of set operations. Such a special knowledge as to how the set is to be used enables us to design specialised data structures to achieve greater efficiency. In this chapter we will study four such restricted set data types with specialised data structures.

6.1 ADT Mapping

A mapping (or a function) M is a mathematical object which is defined from a domain set (D) to a range set (R). This is usually denoted as M : D → R. Mapping M associates a value in R with every element in D. When M associates element r in R to element d in D, we say that d is mapped to r. This is denoted as M(d) = r. Such a mapping is known as a *total mapping* since M is defined for every element of D. Other mappings, where for some elements of D there are no associated values in R, are known as *partial mappings*. For instance, the mapping 'age' is total whereas the mapping 'definition' (see below) is partial. This is because every person has an age, but some identifiers may not have a definition.

   Age: Person → {1. .99}   Age maps a person to his/her age

   Definition: Identifier → Expression   Definition maps an identifier
   to an expression

The latter example is taken from the case study of chapter 3. For each identifier (a letter) a corresponding expression is defined. Here the domain set 'identifier' refers to the set of all possible identifiers. In our example, this set was {'A', 'B', . . ., 'Z'}. A more rigorous definition of a mapping M can be given as a set of pairs. Thus

   M ⊂ D * R

that is, M is a subset of D * R, the cartesian product of sets D and R defined as

   D * R ≡ {(x,y) | x ∈ D and y ∈ R}

The only constraint is that for a given x ∈ D, there can be a maximum of one element y ∈ R such that (x, y) ∈ M. Thus, D * R is the set of all possible ordered pairs of elements and M is a subset of it satisfying this constraint. For example, the mapping 'age' could be:

158

Age ≡ {(Fred, 25), (Joe, 38), (John, 36)}

Now, the abstract data type mapping can be defined as a set of pairs (a binary relation) with the above constraint and the following operations:

procedure initialise (**var** M : mapping);
procedure insert (**var** M : mapping; x : type_D; y : type_R);
procedure delete (var M : mapping; x : type_D);
function value_of (M : mapping; x : type_D) : type_R;

Initialise(M) returns an empty mapping M (set of pairs)
Insert(M, x, y) asserts that (x, y) ∈ M; that is, M(x) = y
Delete(M, x) deletes the pair (x, y) from M (for some value y)
Value_of(M, x) returns y when (x, y) ∈ M (M(x) = y)

Like other sets, simple operations like empty and size may also be defined. Type_D and type_R refer to the type of elements in D and R respectively. Effective implementations of mappings largely depend on the domain set D.

List Implementation of Mappings

The pairs of a mapping can be implemented as a list of pairs where

type pair = **record**
　　　　　　element : type_D;
　　　　　　value　 : type_R
　　　　end;

Pairs p where (p.element, p.value) is a member of the mapping will be organised as a linear list. Insert and delete operations are the usual list operations. Value_of operation should search the list for a pair p with p.element = x and then p.value will be returned. These operations therefore need a maximum time of $O(n)$.

Tree Implementation of Mappings

Since the domain set D is usually ordered, a binary search tree of pairs organised on the basis of the element component of pairs can be used to store the pairs of a mapping. Operations are very similar to the usual binary trees and the average time complexity is $O(\log n)$. The above implementations are general and can cope with any ordered domain set.

Vector Implementation of Mappings

If the domain set is very small and a subset of a small universal set, then an array (a vector) of pointers can be used to represent a mapping. For instance, the mapping 'definition' has a universal domain set U = {'A', 'B', . . ., 'Z'} which is very small, and therefore

type pexpr : ↑expression; (∗ pointer to an expression ∗)
definition : **array** ['A'. .'Z'] of pexpr;

may be used. If a particular identifier (a letter) does not yet have a definition, a nil pointer should be stored in the array. This is very similar to the bit-map representation of sets. It can easily be seen that this implementation gives a time of $O(1)$ for each operation. The extra storage required is $O(N)$ pointers where N is the size of the universal set. This implementation was also discussed at the end of section 5.2.

Unfortunately, this implementation cannot be applied to mappings where the domain set is a subset of a very large universal set. For instance, the mapping 'age' can be represented as

age : **array** [1. .N] **of** integer;

where N is the size of the universal set of person names. Assuming all names are 5-letter strings, N would be 26^5 — that is, approximately 10^7. In practice, however, any set of persons will have far fewer names than 10 million (possibly in the region of thousands). This implementation therefore indicates that $O(10^7)$ of extra storage is required whereas only $O(1000)$ is actually needed. Such a waste of storage cannot be tolerated. For these kinds of domains we shall describe an interesting data structure which does not demand an excessive amount of storage and gives an average time complexity of $O(1)$. Before doing so, let us look at another data type, namely 'dictionary', which has a close affinity to mappings.

ADT Dictionary

The ADT dictionary is a set with the following operations:

  initialise
  insert
  delete
  member

Implementation of dictionaries using lists and trees yields a time requirement of $O(n)$ and $O(\log n)$ respectively. A dictionary may be thought of as a special case of a mapping where the range set contains only two elements {exist, not_exist}. Thus, the value of an element is its existence in the set. As we have discussed before, if the domain set is a subset of a small universal set, then a bit_vector implementation can be used. However, for general and large domain sets, this implementation would waste large amounts of storage. In the next section we describe the hashing technique which overcomes these problems.

6.2 Hashing (Key Transformation)

Any mapping M : D → R can be easily transformed into a mapping H : D → L where L is a set of locations (or addresses). For a given element d in D, then H(d) will be the location where M(d) is stored. Thus, a fast implementation of H would lead to an efficient implementation of M.

For large and fragmented domain sets D, the simple array implementation of H is not satisfactory. The hashing technique assigns a pseudo-random function to H, such that if M(d) = v then H(d) = L, and (d, v) is stored in location L. The storage for these pairs is a table (an array) where each entry in the table corresponds to a location. This table is known as a *hashing table*. Let us assume that the maximum size of the set 'person' in the mapping 'age' is 16. The corresponding hash table for the example of section 6.1 is given in figure 6.1. The set

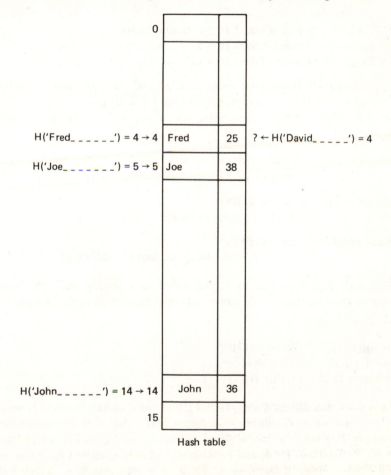

Figure 6.1 Hash table for the mapping 'age'.

location is the set of indices in the array — that is, $\{0. .15\}$. So, mapping H maps every name to a location in the table where that element and its value (from mapping M) will be stored.

H is a pseudo-random function which must be easy and fast to compute. For instance, let us define H to be the following function:

```
function H(n : name) : integer;
  begin
    (* Choose the last letter of the
       name n (non-space char), say L *)
    H := ord(L) mod 16
  end;
```

Thus

$$H(\text{'Fred------'}) = ord(\text{'d'}) \textbf{ mod } 16 = 100 \textbf{ mod } 16 = 4$$
$$H(\text{'Joe-------'}) = ord(\text{'e'}) \textbf{ mod } 16 = 5$$
$$H(\text{'John------'}) = ord(\text{'n'}) \textbf{ mod } 16 = 14$$

It is assumed that ASCII character codes are used, and the type 'name' is a string of 10 characters (see program mapping_age of section 6.2.1).

Using this simple strategy, the mapping operations can be easily implemented as

    insert(M,x,y)   : compute H(x)
                       store (x,y) at location H(x) in the table

    delete(M,x)   : compute H(x)
                       delete the entry at location H(x)

    value_of(M,x)   : compute H(x)
                       return y, where H(x) contains the pair (x,y).

The dictionary data type can be implemented in a similar way. The single difference is that the hash table stores only the element from the domain set. Thus

    insert(D,x)   : store x at H(x)
    delete(D,x)   : clear H(x)
    member(D,x)   : return (H(x) = x)

The major drawback of this strategy is that the pseudo-random function H would not yield unique values for different elements of D — that is, if it is not a surjective mapping. For instance, in the above example, suppose we insert age('David -----') = 40, H('David-----') = 4, and location 4 is already occupied by 'Fred------'. Here H('Fred------') = H('David------'). These two elements have *collided* (or *clashed*) and a *collision handling* procedure is needed to resolve the situation.

It is important to note that any random function is highly likely to be a many-to-one function, so that collisions such as that above are unavoidable. As a classic example, the random mapping 'date-of-birth : person → {1. .365}' is *not likely* to be one-to-one, even for a set of 23 people or more. This is known as the *birthday paradox* and it can be shown that for 23 or more people, the probability of having two people with the same birthday is greater than 0.5.

Therefore, for hashing to be a practical solution to the implementation of mappings, two issues must be dealt with:

(1) collision handling
(2) choice of hashing function H.

The collison handling is primarily resolved by two different methods: closed and open hashing. We shall first discuss these two methods and then, in section 6.2.4, we shall examine a variety of hash functions.

6.2.1 Open Hashing

Let us assume that in the mapping M, the elements $d_1, d_2, \ldots d_k$ have collided, that is

$$H(d_1) = H(d_2) = \ldots = H(d_k)$$

A simple way to resolve these clashes is to store a list of pairs at location $H(d_1)$ and store all the pairs $(d_1, v_1), (d_2, v_2), \ldots$, and (d_k, v_k) in that list where v_i is the value of d_i. Figure 6.2 shows this list of pairs. It is assumed that the hash table is of size N. The table (array) contains one pair and the rest of the collided pairs are stored outside the table in an *overflow area*. Each table entry contains a pointer to a list in the overflow area.

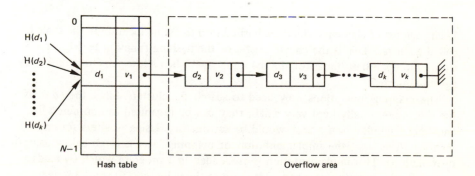

Figure 6.2 Resolving collisions in open hashing.

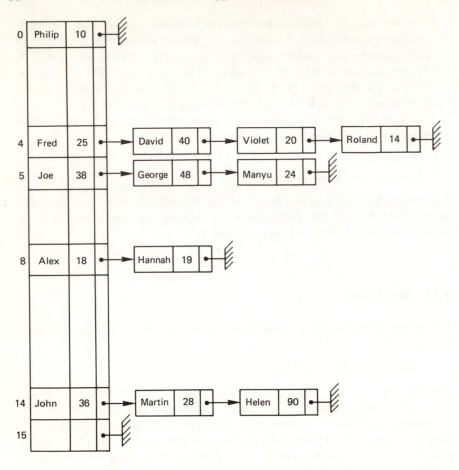

Figure 6.3 Open hash table for the mapping 'age'.

Each such list of elements which are hashed into the same location in the table is called a *cluster*. This is the characteristic of the hashing function H. Figure 6.3 shows the open hash table for the mapping 'age' when a few pairs of names and ages are inserted.

The mapping operations now need to search the clusters. Since the sizes of these lists are usually kept very small, they can be organised as unordered lists and thus a simple linear search would be satisfactory. Using pointers for implementing these lists, the implementation of mapping 'age' is given below using open hashing. This implementation is embedded in a program which (i) reads a sequence of names and ages and adds them to the mapping, (ii) reads a sequence of names and finds their ages, (iii) reads a few names and deletes them from the mapping and finally (iv) reads another list of names to enquire about their ages.

```pascal
program mapping_age (input, output);
  const
    Nminus1 = 15;
    (* Nminus 1 = N-1 where N is the size of the table *)
  type
    index = 0..Nminus1;
    str = packed array[1..10] of char;
    age_type = 1..99;
    domain = str;
    range = age_type;
    entryptr = ↑entry;
    entry = record     (* type of the cells of overflow lists *)
              element : domain;
              value : range;
              next : entryptr
            end;
    hashing = array[index] of entry;
    mapping = hashing;
  var
    age : mapping;
    (* age maps a string(name) to a number 1..99(age) *)
    name : str;
    name_age : age_type;
    error_stat : boolean;

  procedure error;
    (* an error is raised when :-                  *)
    (* insert(M,x,y) : x exists  already           *)
    (* delete(M,x) : x is not in the hash table    *)
    (* value_of(M,x) : x is not in the hash table *)
    begin
      writeln('There has been as error');
      error_stat := true
    end; (* of error *)

  function H (x : domain) : index;
    var
      i : integer;
      ch : char;
    begin
      i := 1;
      while (i < 10) and (x[i] <> ' ') do
          i := i + 1;
      if i = 10 then
        if x[10] = ' ' then
          ch := x[9]
        else
          ch := x[10]
      else
        ch := x[i - 1];
        (* ch is the last character of x *)
      H := ord(ch) mod (Nminus1 + 1)
    end;

  procedure create (var elem : domain);
    begin
      elem := '          '       (* 10 spaces *)
    end; (* of create *)
```

```
function empty_element (element : domain) : boolean;
  begin
    empty_element := (element = '              ')
  end; (* of empty_element *)

procedure initialise (var M : mapping);
  var
    i : index; elem : domain;
  begin
    create(elem);
    for i := 0 to Nminus1 do
      M[i].element := elem
  end; (* of initialise *)

procedure insert (var M : mapping; x : domain; y : range);
  var
    i : index;
    p : entryptr;
  begin
    i := H(x);
    if empty_element(M[i].element) then
      begin
        M[i].element := x;
        M[i].value := y;
        M[i].next := nil
      end
    else if M[i].element = x then
           error    (* x is already in the hash table *)
         else
           begin    (* add to the list of primary cluster *)
             new(p);
             p↑.element := x;
             p↑.value := y;
             p↑.next := M[i].next;
             M[i].next := p
           end
  end; (* of insert *)

procedure delete (var M : mapping; x : str);
  var
    i : index; elem : domain;
    p : entryptr;

  procedure del (var p : entryptr);
    (* search the linked list to find and delete x *)
    var
      q : entryptr;
    begin
      if p = nil then
         error
      else if p↑.element = x then
             begin
               q := p;
               p := p↑.next;
               dispose(q)
             end
           else
             del(p↑.next)
    end;  (* of del *)
```

```
   begin   (* of delete *)
     create(elem);
     i := H(x);
     if empty_element(M[i].element) then (* the list is empty *)
       error
     else if M[i].element = x then
             if M[i].next = nil then
               M[i].element := elem
             else
               begin
                 p := M[i].next;
                 M[i].element := p↑.element;
                 M[i].value := p↑.value;
                 M[i].next := p↑.next;
                 dispose(p)
               end
          else
             del(M[i].next)
   end;   (* of delete *)

function value_of (M : mapping; x : domain) : range;
   var
     i : index;
     p : entryptr;
     found : boolean;
   begin
     i := H(x);
     if empty_element(M[i].element) then
       error
     else if M[i].element = x then
             value_of := M[i].value
          else
             begin
               p := M[i].next;
               found := false;
               (* scan the linked list to find x *)
               while (p <> nil) and not found do
                 if p↑.element = x then
                   found := true
                 else
                   p := p↑.next;
               if found then
                 value_of := p↑.value
               else
                 error
             end
   end; (* value_of *)

procedure read_name (var name : str);
   (* read a name up to 10 characters *)
   var
     i : integer; ch : char;
   begin
     create(name);
     i := 1;
     while (input↑ = ' ') do
       get(input);
     while (input↑ <> ' ') and (i <= 10) do
       begin
```

```
              read(ch);
              name[i] := ch;
              i := i + 1
          end
      end; (* of read_name *)

begin (* Main program *)
  initialise(age);
  (* read a sequence of names and their ages terminated by a $ *)
  writeln('Type a name and an age on one line (or a $ to terminate)');
  read_name(name);
  while name <> '$             ' do
    begin
      readln(name_age);
      insert(age, name, name_age);
      writeln('Type a name and an age on one line (or a $)');
      read_name(name)
    end;

  readln;
  writeln('Type a name on one line to be interrogated (or a $ to terminate)
  read_name(name);
  readln;
  while name <> '$             ' do
    begin
      name_age := value_of(age, name);
      if not error_stat then
        writeln('Age of ', name, ' is ', name_age);
      writeln('Type a name on one line to be interrogated (or a $)');
      read_name(name);
      readln;
      error_stat := false
    end;

  (* few names are to be deleted *)
  writeln('Type a name on one line to be deleted (or a $ to terminate)');
  read_name(name);
  readln;
  while name <> '$             ' do
    begin
      delete(age,name);
      writeln('Type a name on one line to be deleted (or a $)');
      read_name(name);
      readln
    end;

  writeln('Type a name on one line to be interrogated (or a $)');
  read_name(name);
  readln;
  while name <> '$             ' do
    begin
      name_age := value_of(age,name);
      if not error_stat then
        writeln('Age of ',name,' is ',name_age);
      writeln('Type a name on one line to be interrogated (or a $)');
      read_name(name);
      readln;
```

```
        error_stat := false
    end
end.
```

From the above program, it is evident that the time complexity of a mapping operation is proportional to the size of a cluster (a list) which must be searched. In the best case, this would be $O(1)$ if the size is 1. To analyse the average case we must assume that the hashing function is truly random and uniform. This is to say that for any element d, $H(d)$ will be any location 0 to $N-1$ with equal probability. Therefore, in a table of size N, if the number of the elements (pairs) in the tables is M, then the average size of a list will be M/N. To keep the efficiency of the mapping operations, this ratio should be kept very small (usually less than 3 or 4). If this ratio exceeds a given threshold, the hash table should be re-organised with a table twice as large (size $2N$).

A final comment should be made about the storage efficiency of open hashing. In the above solution, for each table entry, space is allocated for at least one pair. If the domain elements are large structures, then it is best to keep only a pointer in the main table and store all the elements in the overflow area (see exercise **6.2**).

6.2.2 Closed Hashing (Rehash)

In closed hashing no overflow area is used. All the pairs are stored in the same hash table. Consider the case when two elements d_1 and d_2 have collided; that is, $H(d_1) = H(d_2) = i$ (say). Let us assume that d_1 is stored at location $H(d_1)$. When hashing d_2, location $H(d_2)$ is occupied. So, a second hashing function H_1 is used to map d_2 to a different location. If, however, $H_1(d_2)$ is also occupied, another hash function $H_2(d_2)$ is examined, and so on. In this technique a sequence of hashing functions is required. These functions are known as *rehash functions*. Naturally, in a table of size N, a maximum of $N-1$ rehash functions should be sufficient to examine the entire table locations. If we consider H as H_0, then this sequence of rehash functions is

$$H_0, H_1, H_2, \ldots, H_{N-1}$$

It is obvious that the same sequence of rehash functions must be used in all the operations.

As an example of a simple rehash strategy, let us define H_i as

$$H_i = (H_{i-1} + 3) \bmod 16$$

The hash table for the above example after the elements Philip, Fred, David, Joe, John, George, Violet, Alex, Hannah, Helen and Martin are inserted in that order will be filled up as shown below. Figure 6.4 shows the final result.

H('Philip----') = 0   O.K.
H('Fred-------') = 4   O.K.
H('David-----') = 4   Collision — Try H_1('David-----') = 4+3 = 7
                                 O.K.

H('Joe-------') = 5   O.K.
H('John------') = 14   O.K.
H('George----') = 5   Collision — Try H_1('George----') = 5+3 = 8
                                 O.K.

H('Violet----') = 4   Collision — Try H_1('Violet----') = 4+3 = 7
                                 Collision — Try H_2('Violet----') = 7+3 = 10
                                 O.K.

H('Alex------') = 8   Collision — Try H_1('Alex------') = 8+3 = 11
                                 O.K.

H('Hannah----') = 8   Collision — Try H_1('Hannah----') = 8+3 = 11
                                 Collision — Try H_2('Hannah----') = 11+3 = 14
                                 Collision — Try H_3('Hannah----') = (14+3) mod 16
                                 O.K.

H('Helen-----') = 14   Collision — Try H_1('Helen-----') = 1
                                 Collision — Try H_2('Helen-----') = 4
                                 Collision — Try H_3('Helen-----') = 7
                                 Collision — Try H_4('Helen-----') = 10
                                 Collision — Try H_5('Helen-----') = 13
                                 O.K.

H('Martin----') = 14   Collision — Try H_1('Martin----') = 1
                                 Collision — Try H_2('Martin----') = 4
                                 Collision — Try H_3('Martin----') = 7
                                 Collision — Try H_4('Martin----') = 10
                                 Collision — Try H_5('Martin----') = 13
                                 Collision — Try H_6('Martin----') = 0
                                 Collision — Try H_7('Martin----') = 3
                                 O.K.

For every rehash function, the content of that location needs to be checked. Such a comparison is known as a *probe* into the hash table. Notice that the earlier elements need one or two probes, whereas the later ones need up to seven probes. Naturally, the efficiency of mapping operations is directly proportional to the number of probes needed.

All the elements of a *cluster* cause collisions in a closed hash table. Such clusters are consequences of the hashing function H and if they are getting large (in size), the choice of the hashing function should be revised (see section 6.2.4).

0	Philip	10
1	Hannah	19
2		
3	Martin	28
4	Fred	25
5	Joe	38
6		
7	David	40
8	George	48
9		
10	Violet	20
11	Alex	18
12		
13	Helen	90
14	John	36
15		

Figure 6.4   Closed hash table for the mapping 'age'.

This kind of clustering belongs to a class known as *secondary clusterings*. To be precise, secondary clusterings are caused by elements such that $H_i(k) = H_i(k')$.

A second phenomenon, however, occurs where two or more clusters are merged together to form a larger cluster. Such new larger clusters are known as *primary clusters*. More precisely, primary clusters occur when $H_i(k) = H_j(k')$ where $j \neq i$. In the example of figure 6.4 the secondary clusters {Joe, George}, {Alex, Hannah} and {John, Martin, Helen} are joined together in a primary cluster, and thus insertion of 'Martin----' requires seven probes. To minimise the number of probes for the mapping operations, the formation of such primary clusters should be avoided (as far as possible). We shall come back to this issue in the next two sections. Before that, one major point should be emphasised. In a linear rehash where $H_i = H_{i-1} + C$, the value of the constant C is crucial to the correct operation of hashing. For the sequence of rehash values $H_0, H_1, H_2, \ldots,$ H_{N-1} to scan the entire hashing table, this sequence must be a permutation of the set of locations $\{0, 1, 2, \ldots, N-1\}$. To ensure this, N and C must be chosen to be *co-prime*; that is, they should not have a common factor. In our example $N=16$ and $C=3$ are co-prime (see exercise **6.4**). A simpler solution which is often adopted is to choose N to be a prime number.

A final point regarding the delete operation is that, when deleting an element, its location cannot be marked as an *empty* location (as in open hashing). Otherwise, this would break the chain of locations of a cluster and make some locations unaccessible to the value_of and insert operations. Such locations should be marked with a distinguishing marker 'deleted'. With this in mind, the mapping operations can be summarised as

```
insert(M, x, y) :
   i := 0
   while (location H_{i mod N}(x) is full) and (i < N) do
     i := i + 1;
   if i = N error (* table in full *)
   else
     (* now H_i is either 'empty' or 'deleted' *)
     (* insert (x,y) in location H_i(x) *)


delete(M,x)  :
   (i)   i := 0; done := false;
   (ii)  repeat
           if H_i(x) contains (x, y) for some y then
             mark that location as 'deleted' and set done to true
           else if H_i(x) is not empty then
             begin
               i := i + 1;
               if i = N then error (* not found *)
             end
         until done;


value_of(M,x)  :
   i := 0;
   while location H_{i mod N}(x) is not empty and (i < N) do
     if H_i(x) contains (x, y) for some y then return y
     else (* either location H_i(x) is 'deleted' or it contains
             a pair (a, b) such that x <> a *)
       i := i + 1;
   if i = N then error;
```

6.2.3 Rehash Strategies

In section 6.2.2 we discussed linear rehashing. To minimise the formation of large primary clusters, we first observe the following two points:

(1) The linear rehash H_i increments H_{i-1} by C irrespective of i. For instance, consider the case where

$$H_0(d_1) = \alpha$$
$$H_0(d_2) = \alpha \qquad H_1(d_2) = \alpha + C$$
$$H_0(d_3) = \alpha \qquad H_1(d_3) = \alpha + C \qquad H_2(d_3) = \alpha + 2C$$

Then

$$H_0(d_4) = \alpha + C \quad \text{and} \quad H_1(d_4) = \alpha + 2C \text{ etc.}$$

The cluster (d_1, d_2, d_3) follows locations α, $\alpha + C$ and $\alpha + 2C$. The cluster $(d_4, \ldots)$ follows locations $\alpha + C$, $\alpha + 2C$ etc. and therefore merges with the first cluster.

(2) The linear rehash H_i is independent of the element being hashed. In the above example

$$H_0(d_2) = \alpha \qquad H_1(d_2) = \alpha + C$$
$$H_0(d_3) = \alpha \qquad H_1(d_3) = \alpha + C \ldots$$

Although d_2 and d_3 are distinct, the sequence of rehash values is not.

Therefore, to improve linear rehashing we should try to eliminate the effects of (1) and (2) above.

Random Rehashing

To combat (1) above, the best technique is to use a sequence of random numbers as increments. That is

$$H_i = (H_{i-1} + R_i) \bmod N$$

where R_i is a sequence of $n - 1$ random numbers. To ensure correct operation, this sequence must be a random permutation of numbers $1 .. N-1$. Therefore, this technique requires a random number generator. Such a random number generator can be developed which, on demand, would give the next element in the sequence. To ensure that the same random sequence is generated each time, the generator should be given the same initial starting value. Special care must be taken to ensure that the random numbers generated do not have duplicates.

Double Hashing

The ultimate solution to combat (1) and (2) above would be for the rehash function to be a function of three arguments i, its previous rehash function $H_{i-1}(i)$ and the element being hashed d. For instance, we may define H_i as

$$H_i(d) = (H_{i-1}(d) + F(i,d)) \bmod N$$

Thus, the rehash $H_i(d)$ is a function of $H_{i-1}(d)$ and $F(i, d)$ which is a second function of i and d. In effect, two hashing functions $H_0(h)$ and $F(i, d)$ are being used. This strategy should break up most chains of clusters. However, this means that two hashing functions must be evaluated each time, which may cause an overhead (in time). In practice, therefore, a compromise should be reached between a fast rehash strategy and minimisation of the primary clusterings in the table.

An important factor affecting the efficiency of hashing which has been left out so far is the *load factor* of the hash table. Load factor is the ratio M/N where M is the number of elements actually in the hash table of size N. Theoretical and experimental results show that when the load factor approaches 100 per cent, the efficiency of hashing operations deteriorates dramatically. This is because, as the table gets full, nothing can stop the formation of large primary clusters.

As a practical guideline, the table should never be allowed to be more than 90 per cent full. If the load factor exceeds 90 per cent, the table should be re-organised with a larger size (new size = $2N$).

6.2.4 Hashing Functions

In the example of section 6.2, we chose a very simple hashing function H. In this section we shall analyse that function and give some guidelines on constructing good hashing functions.

Primarily, a hashing function must have two properties:

(1) It should be a simple function, and consequently it should be fast to evaluate.
(2) It should distribute the elements in the hash table uniformly.

The first criterion is necessary because each mapping operation needs to evaluate this function. The second criterion ensures that clusters are kept to a minimum size.

Although the function H of section 6.2 satisfies (1) above, it distributes the elements very unevenly. Since the function H depends only on one character of the elements, many elements are clustered together. To overcome this

a hash function must depend on all parts of its argument.

A very good and often-used function is the 'middle_of_square' function. This finds the square of a given argument and then extracts the middle bits of that number. It is assumed that the argument is a number. If character strings are used as arguments, binary representation of the strings should be used. This general technique is known as *extraction*; that is, extracting a few bits from an element or from a function of an element.

A second strategy for designing a hash function is by *compression*. The elements are broken into a few components, then the components are compressed together. For instance, character strings can be broken up into individual characters, then the codes of these can be combined to give a final value. The usual way to do this is to 'exclusive_or' these components. For example, if ASCII characters codes are used:

$$
\begin{array}{ll}
\text{H('David----')}: & 01100100 \oplus \\
 & 01100001 \oplus \\
 & 01110010 \oplus \\
 & 01101001 \oplus \\
 & \underline{01100100} \\
 & 01111010 = 7 \times 16 + 10
\end{array}
$$

$$
\text{H('David----')} = (7 \times 16 + 10) \bmod 16 = 10
$$

This procedure ensures that all the bits are used in the hashing function. The $\oplus$ above is the 'exclusive_or' operation. In fact, addition can also be used, but $\oplus$ does not produce any carry and so does not lose any information. Compression is also a good technique for reducing the lengths of elements to be hashed. Thus, in the above example, $7 \times 16 + 10$ could be used as an argument of another hashing function.

If we are using the above technique for mapping English words (in constructing a dictionary, say), then:

$$
\text{H('steal----')} = \text{H('stale----')} \text{ etc.}
$$

Thus, any permutation of a set of components yields similar results. To stop such biased behaviour, we can take into account the order of characters. For example, we may 'rotate' the result before doing another 'exclusive_or' operation. If $d = s_1 s_2 \ldots s_k$ then

$$
\text{H}(d) = ((s_1 \oplus s_2)' \oplus s_3)' \oplus \ldots
$$

where $(\ )'$ is a rotate operation (bitwise).

Two widespread classes of hashing functions are division and multiplication functions.

In the *division* method, function H is simply defined as

$$
\text{H}(d) = d \bmod N
$$

That is, $\text{H}(d)$ is the remainder of the division d/N. This function is fast to compute. However, N must be chosen carefully to ensure its uniformity. For instance, if N is a power of 2, then $\text{H}(d)$ would return the m most significant bits of d where $N = 2^m$. This means that $\text{H}(d)$ does not depend on all the bits. Similarly, if N is a power of k (for some k), the hash values will not depend on all the bits of the elements. Even if N has a factor which is a power of k, similar anomalies will arise. In the limit, N should be a prime number if H is to depend on all the bits of its argument (irrespective of N). Note that, in this method, the size of the table cannot be any arbitrary number.

In the *multiplication* method, function H is defined as

$$H(d) = \text{trunc}\ (N \times (d \times \theta \bmod 1))$$

where θ is a real number $0 < \theta < 1$. $(d \times \theta \bmod 1)$ returns the fractional part of $(d \times \theta)$. Thus

$$0 \leqslant d \times \theta \bmod 1 < 1$$

$$0 \leqslant N \times (d \times \theta \bmod 1) < N$$

Therefore $0 \leqslant H(d) \leqslant N - 1$. The advantage of this technique is that, firstly, N may be any arbitrary number. Secondly, certain values of θ result in functions which theoretically have good random behaviour. For example, if θ is the *golden ratio* 0.618033, then simple progressions in the input are broken uniformly (for example, part1, part2, part3 etc.).

6.3 Priority Queues

The second special ADT based on sets is priority queues. In chapter 3 we defined a stack as a Last In First Out (LIFO) structure and a queue as a First In First Out (FIFO) structure. By analogy with stacks and queues, we define a priority queue to be Smallest In First Out (SIFO) structure (or Greatest In First Out — GIFO — by symmetry). A priority queue is very much like a queue, with the difference that the elements have priorities associated with them. Each time an element is to be deleted, the one with the highest priority is chosen. The linear relationship 'next' and 'previous' does exist among the elements in a dynamic sense, but it is usually reflected in the way priority queues are used. Therefore, a priority queue can be thought of as a set of elements which are ordered according to their priorities. We define a priority queue as an ADT set with the following operations:

> **procedure** initialise (**var** Q : priority_queue);
> **procedure** insert (**var** Q : priority_queue; x : elemtype);
> **procedure** delete_min (**var** Q : priority_queue; **var** x : elemtype);

The two functions initialise and insert are as for general sets. The delete_min procedure finds the element with minimum priority, assigns it to x and finally deletes it from the set. The generic abstract data type 'elemtype' should, among other operations, provide the following operation:

> **function** priority (x : elemtype) : integer;

As usual, it is assumed that the type 'elemtype' supports operations for testing (for equality, greater and less than) and assignment.

Primarily, priority queues are used for simulation purposes where objects enter a queue to be processed. When an agent (a processor) becomes free, the object with maximum urgency (highest priority) is removed from the priority queue to be processed. For instance, the behaviour of a hospital waiting room resembles that of a priority queue. As patients arrive, the urgent cases are dealt with first. Priority queues are often used in computer systems for a variety of purposes. One such application is when many user-programs and system-programs are processed in a time-sharing operating system. As there is only one computer, its time is shared between different programs. Each time the computer finishes one program, it will pick up another program with the highest priority to run. It is obvious that some programs such as a disc manager are more urgent than others. Also, to reduce response times, short programs may be preferred over large programs etc. Thus, each program is assigned a priority before being added to a priority queue of programs.

Finally, priority queues are used for developing efficient algorithms for certain problems in computer science, numerical analysis etc. In chapter 7 we shall discuss an interesting application of priority queues.

Implementation of Priority Queues

Using ordered or unordered lists, the nominal time for insert and delete_min would be $O(n)$. Thus, a mix of n insert and delete_min operations needs a time of $O(n^2)$. This makes list implementation suitable only for small values of n. Binary search trees organised according to the priorities of elements can be used to give an average time of $O(\log n)$ for priority queue operations. This is fine, except that for a priority queue of length n, $O(2n)$ extra storage is needed for the pointers in binary search trees. Note also that linked list implementations of priority queues need $O(n)$ extra storage for pointers. Now, we shall look at a specialised implementation which would give a guaranteed worst-case time of $O(\log n)$ without using extra storage. We shall achieve this in two stages: first by implementing a priority queue using a partially ordered binary tree, and then by implementing a partially ordered tree using a heap.

6.3.1 *Partially Ordered Binary Trees*

A binary search tree is *fully* ordered. That is to say, for each node, its left sub-tree contains smaller elements and its right subtree contains larger elements than the one in that node. We shall define a partially ordered binary tree (POBT) as an ADT binary tree with the following two constraints:

(1) A partially ordered binary tree is a *complete* binary tree. A complete binary tree has exactly 2^i nodes at each level i of the tree, except at the lowest level where fewer nodes may exist and the leaves are as far to the left as possible. Therefore, to construct a complete tree, level i should be filled before level $i + 1$ and, at the lowest level, nodes should be added as far to the left as

possible. In other words, a strict top-down, left-to-right approach is used to construct POBTs. Two examples of complete trees are given below. The numbers show the order in which they were added to the tree. Note that the structure of a complete tree with n nodes is unique.

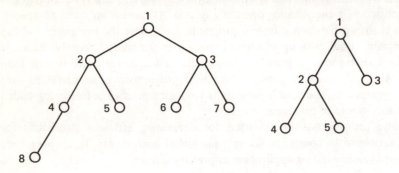

(2) In a partially ordered binary tree, for each node N containing element e, its left and right subtrees contain elements *not less than* e. An example of a POBT is shown below:

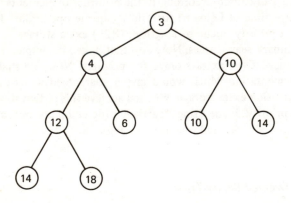

Notice the differences between a POBT and a binary search tree. Element 14 is present in two subtrees of node 3.

According to the definitions above, firstly, the height of a POBT with n nodes is trunc(log n) — that is, the integer part of log n. Secondly, the smallest element must be at the root of the POBT. This follows directly from constraint (2).

Using a POBT we can now define operations insert and delete_min of priority queues.

insert(B,e):   To insert the new element e into POBT B, first we add e to

B to make a complete tree with $n + 1$ nodes (assuming B has n nodes). Definition (1) implies that there is a unique place for e: the node on the lowest level as far to the left as possible. If the lowest level is full, a new level should be started. For instance, adding element 2 to the above example would yield

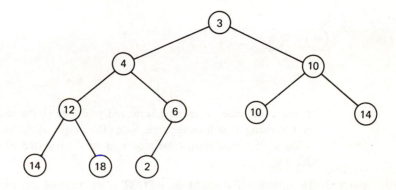

To ensure the invariance of constraint (2), firstly, we observe that only node 6 is affected by adding 2. Therefore, by comparing 2 and its parent 6, it is evident that they should be exchanged:

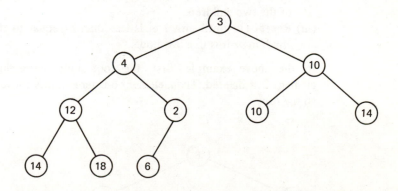

In a similar way, the new node containing element 2 must be compared with its parent node and exhanged, if necessary, until either it is not less than its parent or it is the root of the POBT. In this case, 2 should be replaced with 4 and 3 repeatedly.

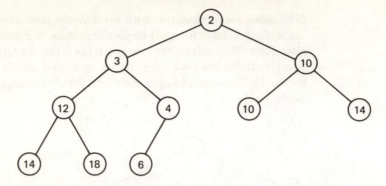

In the worst case, element e is moved gradually to the root by traversing a path of the tree. Since the height of the tree is log *n*, the maximum time requirement for insert(B,e) is O(log *n*).

delete_min(B,e): The minimum element of a **POBT** is the root of the tree. The element at the root should therefore be returned (in e). Deleting the root, however, leaves us with two subtrees. To maintain the constraints, a simple algorithm can be used:

  (i) Temporarily, move the element at the lowest level and as far to the right as possible to the root (say e').
  (ii) Compare e' with its two children nodes and, if e' is smaller than both of them, replace e' with the smaller of the two children.
  (iii) Repeat (ii) until either e' is less than or equal to its children *or* it is in a leaf node.

In the above example, first, the root node containing element 2 is deleted. Then, element 6 is temporarily moved to the root:

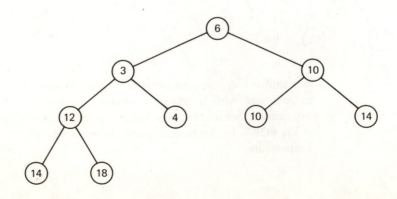

Now, element 6 is pushed down successively to its correct place. First, node 6 is replaced by the minimum of elements 3 and 10:

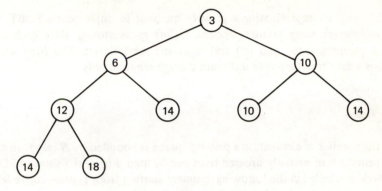

Again, node 6 should be replaced by the minimum of its children — nodes 12 and 4 — that is, node 4, after which it will be in a leaf node. Therefore, the final tree is

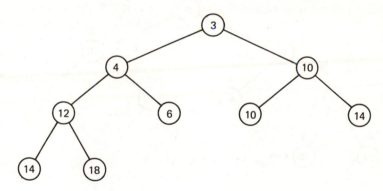

After performing three more delete_min operations, the tree will look like this:

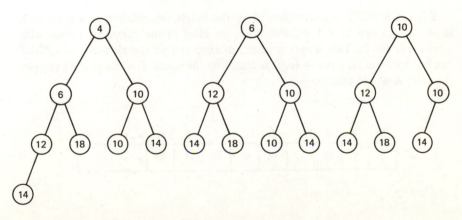

The complexity of delete_min is O(log n) in the worst case since the temporary element at the root will traverse a path to a leaf node (in the worst case).

The next step in implementing a priority queue is to implement a POBT. The usual techniques using pointers require O($3n$) extra storage since each node needs a parent pointer, and left and right subtree pointers. The next section describes a data structure with optimum storage requirements.

6.3.2 Heaps

When the number of elements in a priority queue is bounded by N (say), an array implementation of partially ordered trees can be used. Figure 6.5 shows a POBT. Each node is labelled in the following manner: starting from 1, consecutive labels are given to nodes in a top-down, left-to-right fashion.

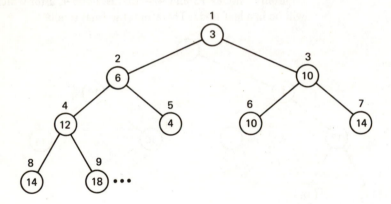

Figure 6.5   Labels in a POBT.

Because a POBT is a complete tree, the labels for children of a node with label i are $2i$ and $2i + 1$. Furthermore, the label of the parent of a node with label i is (i **div** 2). This simple mechanism ensures that the elements of a POBT can be stored in an array A (say) indexed by the labels. For the above example, the array A would be

	1	2	3	4	5	6	7	8	9	• • •	N
A:	3	6	10	12	14	10	14	14	18	• • •	

The children of A[i] are A[$2i$] and A[$2i + 1$]. As the tree was partially ordered, the array satisfies the conditions:

$$\left. \begin{array}{l} A[i] \leqslant A[2i] \\ \\ A[2i] \leqslant A[2i + 1] \end{array} \right\} \text{for } 1 \leqslant i \leqslant (N \text{ div } 2)$$

Such an array structure is known as a *heap*. The operations of partially ordered binary trees can be directly applied to heaps where the only difference is in the way the parent or children of a node are accessed:

```
const N =    ;
type heap = record
              elements : array[1..N] of elemtype;
              eoh : 0..N    (* end of heap *)
           end;
     priority_queue = heap;

procedure initialise(var Q : priority_queue);
   var i : integer;
   begin
     Q.eoh := 0
     for i := 0 to N do Q.elements[i] := elem0;
     (* elem0 is an element of type elemtype *)
   end;  (* of initialise *)

procedure insert(var Q : priority_queue; x : elemtype);
   var i, parent : integer; temp : elemtype;
   begin
     if Q.eoh = N then error        (* Q is full *)
     else
        begin
           i := Q.eoh + 1;
           Q.elements[i] := x; Q.eoh := i;
           parent := i div 2;

           while (i > 1) and  (* the root has not yet reached *)
                 (priority(Q.elements[i])
                      < priority(Q.elements[parent])) do
              begin
                 (* Exchange elements i and parent *)
                 temp := Q.elements[i];
                 Q.elements[i] := Q.elements[parent];
                 Q.elements[parent] := temp;
                 i := parent;
                 parent := i div 2
              end
        end
   end;    (* of insert *)
procedure delete_min(var Q : priority_queue; var x : elemtype);
   var temp : elemtype;
       i,j : integer;
       more : boolean;
   begin
     if Q.eoh = 0 then error        (* Q is empty *)
     else
```

```
begin
    x := Q.elements[1]; (* element with smallest priority *)
    if Q.eoh = 1 then Q.eoh := 0
    else
        begin
            (* move the last element to the root *)
            Q.elements[1] := Q.elements[Q.eoh];
            Q.eoh := Q.eoh - 1;
            (* Now push the first element into the heap *)
            i := 1;
            more := true;
            while (i <= (Q.eoh div 2)) and more do
                begin
                    if (priority(Q.elements[2*i])
                        < priority(Q.elements[2*i+1])) or
                            (2*i = Q.eoh) then
                            j := 2 * i
                    else j := 2*i + 1;
                    (* j is the smaller child of i *)
                    if priority(Q.elements[i]) >
                                priority(Q.elements[j]) then
                        begin
                            (* exchange these two elements *)
                            temp := Q.elements[i];
                            Q.elements[i] := Q.elements[j];
                            Q.elements[j] := temp;
                            i := j
                        end
                    else                    (* stop the process *)
                        more := false
                end    (* of while *)
        end
    end
end;  (* of delete *)
```

6.4 The ADT Relation

In chapter 5, we defined TRIPLE_SET as a set of triples. We can generalise the idea of a TRIPLE_SET to sets of larger objects. Let $D_1, D_2, \ldots, D_n$ be n domain sets. A *tuple* $(a_1, a_2, \ldots .a_n)$ is defined as an ordered sequence of n objects each from a domain D_i, $a_i \in D_i$ for all i.

Using the language of sets:

$$(a_1, a_2, \ldots, a_n) \in D_1 * D_2 * D_3 * \ldots * D_n$$

That is, a tuple is a member of the cartesian product of the sets $D_1, D_2, \ldots, D_n$. Now, a *relation* R is defined to be a set of such tuples. In other words:

$$R \subset D_1 * D_2 * \ldots * D_n$$

Such an R is called an *n*-ary relation.

For $n = 3$, the resulting ternary relation R may also be called a TRIPLE_SET. For $n = 2$, the resulting binary relation R can be thought of as a PAIR_SET. A

PAIR_SET is usually known as a *many-to-many relationship*. In the remainder of this chapter, we shall examine the ADT PAIR_SET and its implementations in some detail. The definition and implementations of PAIR_SETs can then be generalised for *n*-ary relations.

A PAIR_SET P is a set of ordered pairs of elements (a, b) where a is of type elemtype_A and b is of type elemtype_B (these two are generic data types such as elemtype). Therefore P $\subset$ elemtype_A $*$ elemtype_B. For instance, 'enrolment' can be thought of as a PAIR_SET where each enrolment consists of a (student, course) pair. Figure 6.6 shows an instance of the enrolment PAIR_SET.

Student	Course
Fred	CC206
Fred	CC205
Jim	CC205
Joe	CC201
Jim	CC201
Helen	CC206
Claire	CC201

Figure 6.6   An instance of the 'enrolment' PAIR_SET (binary relation).

As an ADT, a PAIR_SET can be defined as a set of ordered pairs with operations:

```
procedure initialise (var P : PAIR_SET);
procedure insert (var P : PAIR_SET; x : pair);
procedure delete (var P : PAIR_SET; x : pair);
procedure delete_A (var P : PAIR_SET; a : elemtype_A);
procedure delete_B (var P : PAIR_SET; b : elemtype_B);
function member (P : PAIR_SET; n : pair) : boolean;
procedure find_A (P : PAIR_SET; a : elemtype_A; procedure proc(x : pair));
procedure find_B (P : PAIR_SET; b : elemtype_B; procedure proc(x : pair));
function empty (P : PAIR_SET) : boolean;
```

The type pair is defined as:

```
type pair = record
                a : elemtype_A;
```

```
            b : elemtype_B;
            other : other_type
        end;
```

The 'other' field allows extra information to be added about a pair.

The only operations which are different from general sets are delete_A, delete_B, find_A and find_B.

delete_A (P,a): All pairs (a,x) ∈ P are deleted
delete_B (P,a): All pairs (x,b) ∈ P are deleted
find_A (P,a,proc): All pairs (a,b) such that (a,b) ∈ P are processed by the procedure proc
find_B (P,b,proc): All pairs (a,b) such that (a,b) ∈ are processed by the procedure proc

In our example above, find_A(enrolment, student, proc) would process all courses for a given student using the procedure proc. Similarly, all students in a course can be processed using find_B(enrolment, course, proc).

6.4.1 Implementation of the ADT PAIR_SET

The easiest implementation of a PAIR_SET is by using a two-dimensional array. A PAIR_SET can be visualised as a set of points in a two-dimensional space. For instance, the example of figure 6.6 would look like this:

The operations find_A and find_B correspond to the marked elements of the rows and columns respectively of this matrix. If the matrix is very *dense* — that is, if a PAIR_SET contains most of the possible pairs of the matrix — then this implementation will be a viable solution. In practice, however, only a limited number of those points are present in a PAIR_SET and therefore matrix implementation would waste a lot of storage for non-used elements of the array. Such matrices are called *sparse* matrices. In the next section, we look at two alternative approaches for representing sparse matrices.

6.4.2 *Multi-list Implementation of PAIR_SETs*

One way to implement a sparse matrix is to store only the non-empty elements of the matrix. Therefore, in the above example, lists of elements in rows or columns can be constructed. To be able to process find_A and find_B operations efficiently, separate lists of rows and columns should be built. In this way, any element would be in two different lists. Since each element may consist of a large information record ('other' field), to eliminate duplication of this piece of data, such lists can be integrated into a *multi-list* structure (see figure 6.7).

In figure 6.7, pairs (a,b), (a,b'), (a',b) and (a'',b) are represented. Each element of a PAIR_SET has at least four pointers. The first two point to the corresponding components of a pair. The third is a pointer to the next pair with the same elemtype_B component, and the last pointer points to the next pair (if any) with the same elemtype_A component. Each element from elemtype_A and elemtype_B has a pointer to point to the beginning of its list. These inter-linked lists represent rows and columns of a sparse matrix. Applying this representation to the enrolment set, we would get the structure of figure 6.8. Note that each element of a PAIR_SET may have extra information (such as the grade of each student in a course).

This structure can be generalised for an *n*-ary relation where each element of the relation belongs to (up to) *n* distinct lists. It should be noted that for an *n*-ary relation, there will be other types of operations, such as search operations, when *k* values ($k \leqslant n$) for *k* domains are known and the values for *n−k* domains are left unspecified. For instance, in the TRIPLE_SET of chapter 5, the operation attribute_of needs values for two domains and returns values from the third domain (see exercise **6.15**). These general types of operations are called *partial match* operations on relations.

6.4.3 *Multi-dimensional Tree Implementation of PAIR_SETs*

The usual binary search trees can be thought of as means of partitioning a one-dimensional search space. For instance, the following tree (page 189) partitions the set of numbers {2, 10, 15, 35, 45, 80} into two sets {2, 10, 15}, {45, 80} and a singleton set {35}.

This scheme can be generalised to search spaces of more than one dimension. For a two-dimensional space − that is, a sparse matrix − trees can be used to

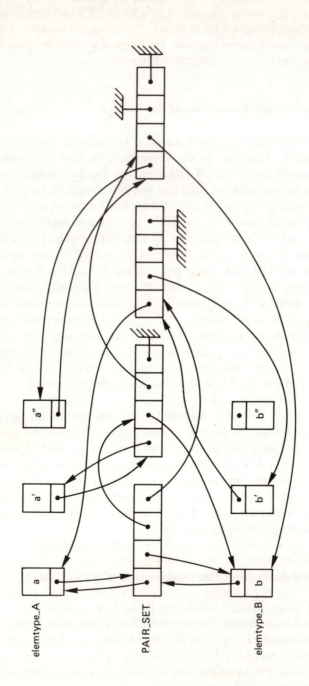

Figure 6.7 Multi-list representation of a PAIR_SET.

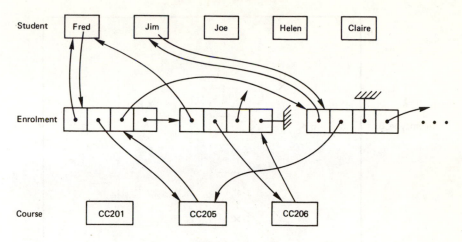

Figure 6.8 Multi-list representation of enrolments.

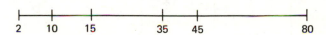

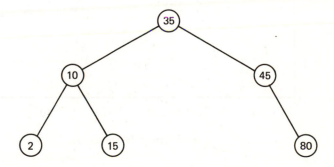

partition the two-dimensional space and consequently reduce (up to a half) the amount of search for find_A and find_B operations.

Let us use a PAIR_SET where the type of components are integers $1 \ .. \ N$. Figure 6.9 shows an instance of such a PAIR_SET (given as a two-dimensional matrix).

Let us take these pairs and construct a tree in the order A, E, G, **B, D, F, C, H, I** and **J**. Initially our tree is empty. Pair A would produce a tree with a root and no children (see figure 6.10(i)). In analogy to the usual binary search trees, the pair A can divide the area into two sub-areas. This partitioning can be done along

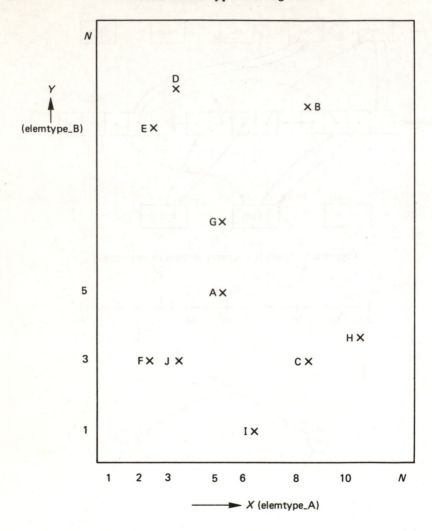

Figure 6.9 A PAIR_SET.

the line where $X = 5$. Thus the area is divided into one area to the left of point A and one area to the right of point A (see figure 6.11). We say that node A has an *X-discriminant*. When adding the second point, E, since the *X*-value of E is less than that of A, it should be added to the left sub-tree of A (figure 6.10(ii)). To ensure that we can process find_A and find_B operations, we now partition the area to the left of A into two sub-areas along the line segment $Y = 9$ ($X < 5$), as in figure 6.11(ii). Thus node E has a *Y-discriminant*. Now, we can add the other points in turn and, each time we move down the tree by one level, the discriminant is changed from X to Y and vice versa. Figure 6.10 shows how the tree

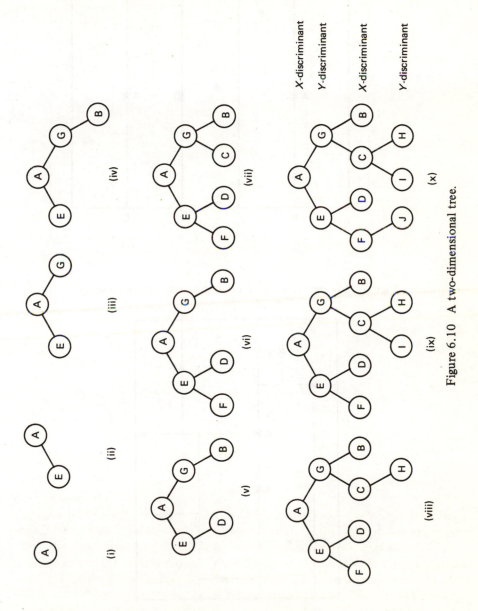

Figure 6.10   A two-dimensional tree.

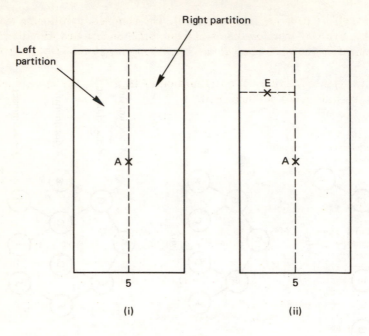

Figure 6.11   Partitioning of a two-dimensional search space.

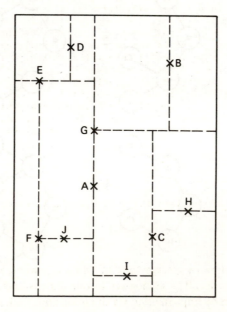

Figure 6.12   Complete partitioning of the two-dimensional space.

grows as all the pairs are inserted into it. The complete partitioning resulting from this tree will therefore be as in figure 6.12. Such a tree with alternating discriminants is known as a *two-dimensional binary search tree* (figure 6.10(x)).

Let us now look briefly at operations on PAIR_SETS. The insert operation follows the strategy we outlined above.

```
type discr = (A,B);
     nodeptr = ↑node;
     node = record
                 element : pair;
                 left,right : nodeptr
            end;
     PAIR_SET = nodeptr;
........
procedure insert (var P : PAIR_SET; x : pair);
  procedure insert_2_tree (var P : PAIR_SET; x : pair; D : discr);
    begin
      if empty(P) then
        begin                      (* create a new node *)
          new(P);
          P↑.element := x;
          P↑.left := nil;  P↑.right := nil
        end
      else
        case D of
          A : if x↑.a < P↑.element.a then
                  insert_2_tree(P↑.left, x, B)
              else if (x↑.a = P↑.element.a)
                         and (x↑.b = P↑.element.b) then
                   error (* the pair x is already in the tree *)
                   else insert_2_tree (P↑.right, x, B);
          B : if x↑.b < P↑.element.b then
                  insert_2_tree(P↑.left, x, A)
              else if (x↑.b = P↑.element.b)
                         and (x↑.a = P↑.element.a) then
                   error (* the pair x is already in the tree *)
                   else insert_2_tree (P↑.right, x, A)
        end
    end;  (* of insert_2_tree *)

  begin (* of insert *)
    insert_2_tree(P, x, A)
    (* assuming that we start the tree using a A discriminator *)
  end;  (* of insert *)
```

This procedure requires an average time of O(log m), where m is the number of pairs in the set (this result is analogous to one-dimensional binary trees). The find_A operation can be implemented via the procedure find_A_2_tree:

```
procedure find_A_2_tree (P: PAIR_SET; a : elemtypeA; procedure
                                      proc(n : pair); D : discr);
   begin
     if not empty(P) then
        begin
          if P↑.element.a = a then proc(P↑.element);
          case D of   (* now search the subtrees for any matches *)
          A : if a >= P↑.element.a then
(* 1 *)         find_A_2_tree (P↑.right, a, proc, B)
              else
                find_A_2_tree (P↑.left, a, proc, B);
          B : begin
(* 2 *)         find_A_2_tree (P↑.left, a, proc, A);
                find_A_2_tree (P↑.right, a, proc, A)
              end
        end;
   end;  (* of find_A_2_tree *)
```

Note that in (* 1 *) only the right or the left subtree should be searched (not both), whereas in (* 2 *) the two subtrees should be searched.

The find_B procedure can be written similarly. The analysis of such multi-dimensional trees is difficult. It can be shown that the find_A and find_B operations need a time of $O(\sqrt{m})$ on average.

Multi-dimensional trees can be constructed for n-ary relations where the discriminator is alternated and repeated every n times. For a partial match in an n-dimensional tree when k values are known, the time complexity can be shown to be

$$O(m^{(n-k)/k})$$

Therefore, these trees do not compare favourably with multi-lists where the responders can be read from a list. However, when n is large and where partial match queries with $k > 1$ are used often, multi-dimensional trees may be better since they only require $O(2m)$ of extra storage for pointers, whereas multi-lists would need $O(n \times m)$ extra pointers. Another advantage of these trees is in *range searching* (see exercise **6.18**).

Exercises

6.1. In open and closed hashing, when the load factor exceeds its threshold, how can re-organisation of the hash table be achieved?

6.2. Modify the open-hash implementation when the hash table contains only a set of pointers. What are the advantages and disadvantages of this scheme?

6.3. Suppose that we have a hash table with 16 buckets (N=16). Show the resulting table if closed hashing with linear rehashing ($H_n \equiv H_{n-1} + 1$) is used after the following numbers are inserted in this order:

1 4 9 16 25 36 49 64 81 100 121

Use the hashing function H(i) = i mod 16.

6.4. What happens if, in exercise **6.3**, the rehash function $H_n \equiv H_{n-1} + 4$ is used? Why is this scheme unsuccessful?

6.5. Define the ADT partially ordered binary tree.

6.6. Using the ADT of question **6.5**, implement the ADT priority queue.

6.7. Insert the following elements into an empty partially ordered binary tree:

10 9 8 7 6 5 4 3 2 1.

6.8. Perform 10 *deletemin* operations on the tree constructed in exercise **6.7**.

6.9. Complete the implementation of a priority queue using a heap structure.

6.10. In the Huffman algorithm of chapter 4, a list was used to maintain a forest of trees. How can the algorithm be improved by using a priority queue? Implement these two alternatives and compare their efficiencies.

6.11. For certain mappings (dictionaries of English words) where the domain contains elements which are *strings* of data objects, a tree structure known as *tries* can be used. For instance, if the elements are character strings, then a dictionary can be represented as a mapping:

Trie : {A, B, C, . . ., Z, $} → Trie

For example, the elements {RING, RIM, RAT, SAD} are represented as

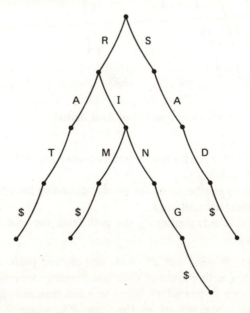

The '$' indicates the end of a word. Such a representation has a time requirement proportional to the *length* of a string.

Formalise a *trie* as an ADT.

6.12. Investigate how tries can be implemented using arrays and/or pointers of Pascal. What are the time/space requirements?

6.13. *Practical assignment*

A linguist is studying the frequency of occurrences of words in pieces of text. He also wishes to gather statistics about prefixes in the language being studied. For this purpose, a particular trie structure can be used to build a dictionary of words. Using the linked list representation of a trie, the set of words {RING, RIM, RAT, RINK, SAFE, SAD} would look like this:

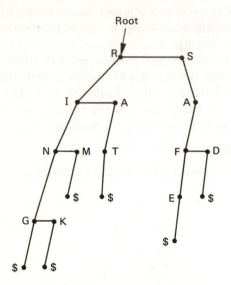

Every node of such a trie consists of four items:

1. A letter or a $.
2. A pointer to the next node for words containing that letter and previous letters (left_child).
3. A pointer to another node for words containing the previous letters but *not* this one (right_sibling).
4. A count of words containing the path from the root up to and including this letter.

It is required to print out all words and *distinct* prefixes in alphabetical order, together with counts of their occurrences. A prefix P is a sequence of one or more letters which occur in more than *one distinct word* (disregarding duplicate words) in the form PX, where X is a non-empty sequence of letters. For instance, for the above set of words the output would be

R–	4
RAT	1
RI–	3
RIM	1
RIN–	2
RiNG	1
RINK	1
SA–	2
SAD	1
SAFE	1

Note that 'S– 2' is not required because it is included in 'SA– 2'. However, if the word 'SIN' were added to the dictionary, the output would include 'S– 3' and 'SIN 1'.

A prefix which is also a complete word should be printed as both. As an example, if the words 'MAN' and 'MANKIND' are added to the dictionary, 'MAN' is not a prefix and the output should be

MA–	2
MAN	1
MANKIND	1

If the word 'MANY' is added the result would contain

MA–	3
MAN	1
MAN–	2
MANKIND	1
MANY	1

You are required to write a program which performs:

1. Reading in a set of words
2. Building a dictionary.
3. Printing out the results in alphabetic order. Printing is easier if an alphabetically sorted trie is constructed.

6.14. Choose suitable data structures and write Pascal code for find_all_courses_for_a_student and find_all_students_for_a_course using a multi-list structure.

6.15. *Practical Assignment: implementation of a general memory structure to represent information*
General symbolic information can be represented in terms of TRIPLEs of identifiers. For example

(john	is-a	student)
(john	likes	mary)
(fred	is-a	student)

```
(fred      hates      john)
(john      earns      6000)
(fred      lives_in   london)
. . . . . . . . .
        etc.
```

We may then wish to set exercises or ask questions such as

- List all the students   (John, Fred)
- Who likes Mary?   (John)
- How much does John earn?   (6000)
- How are John and Fred related? (Fred hates John)
  etc.

The objective of this assignment is to develop an ADT for manipulating *triples*. A triple T is defined as an ordered triplet (object, attribute, value) where each object, attribute or value is an identifier. An identifier is a string of characters up to 20 char. Triples are unique − that is, there cannot be more than one triple with the same object, attribute and value identifier. The operations to be performed on the triples are

procedure insert_triple(object, attribute, value);
   This procedure adds a triple to the TRIPLE_SET.

function verify(object, attribute, value) : boolean;
   A boolean function which determines if a triple exists in the TRIPLE_SET.

function value_of(object, attribute) : LIST (* of identifier *);
   This function returns all identifiers Id such that the triple (object, attribute, Id) exists in the TRIPLE_SET.

function attribute_of(object, value) : LIST (* of identifier *);

function object_of(attribute, value) : LIST (* of identifier *);
   These two are similar to value_of function.

function explore_object(Id) : LIST (* of PAIR *);
   This function returns all the triples with Id as the object field.

function explore_attribute(Id) : LIST (* of PAIR *);

function explore_value(Id) : LIST (* of PAIR *);
   These two are similar to explore_object.

A suitable data structure for triples is

```
identifier = packed array [1..20] of char;
tripleptr = ↑triple;
triple = array [1..3] of record
```

id : identifier;
ptr : tripleptr
end;

For example

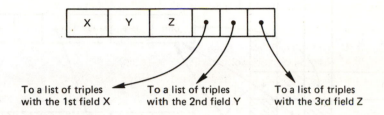

A *set* of identifiers (LEXICAL_TABLE) provides entry into triple structures. The LEXICAL_TABLE has the following structure:

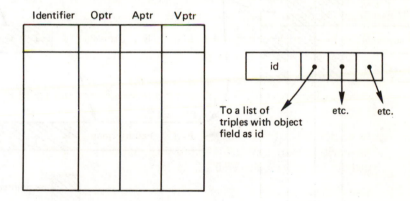

The following operations are supported by LEXICAL_TABLE:

> **procedure** insert_id (item : identifier; opr, apr, vpr : tripleptr);
> **procedure** retrieve(item : identifier; **var** opr, apr, vpr : tripleptr);
> **procedure** delete_id(item : identifier);

Figure 6.13 shows the memory structure after a few triples have been inserted.

(1) Fully define triple, triple_memory and lexical_table as ADTs. What other operations are necessary?
(2) Implement the above ADTs and embed them in a program which reads commands:

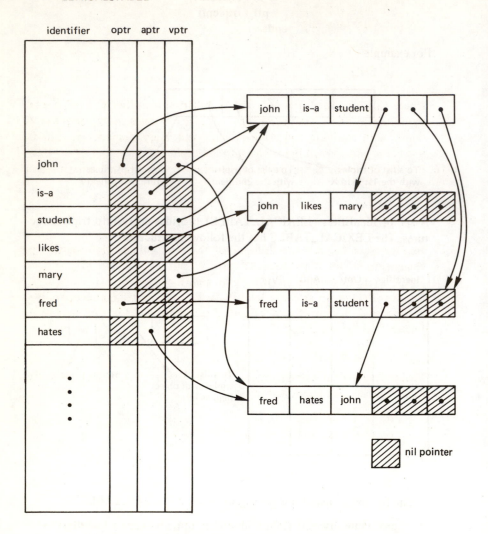

Figure 6.13   Memory structure after a few triplets are added.

INSERT
⟨id⟩ ⟨id⟩ ⟨id⟩
⟨id⟩ ⟨id⟩ ⟨id⟩
  ..
  ..
END

INTERROGATE
⟨arg⟩ ⟨arg⟩ ⟨arg⟩
⟨arg⟩ ⟨arg⟩ ⟨arg⟩
. .

. .
END

where ⟨id⟩ is an identifier and ⟨arg⟩ is either an identifier or a '*' symbol signifying a 'don't care' state. For instance, "* likes mary" requires all triples with attribute and value as 'likes' and 'mary' respectively. The insert command adds all the triples into the triple_memory.

6.16. For the following set of pairs, construct a two-dimensional binary search tree:

$$(5, 5) \quad (7, 3) \quad (2, 8) \quad (5, 7) \quad (5, 1)$$
$$(5, 9) \quad (8, 1) \quad (1, 3) \quad (6, 5).$$

6.17. In exercise **6.16**, can the ordering of the 'discriminant' information be changed to obtain more 'balanced' trees? How?
Devise a scheme for implementing your solution.

6.18. *Range-searching:* In an *n*-ary relation, apart from partial-match operations as discussed in this chapter, another type of operation can be defined known as range-searching. If

$$R \subset D_1 * D_2 * \ldots * D_n$$

then a range search is a request for all tuples $(a_1, a_2, \ldots, a_n) \in R$ such that each value lies in a prescribed range:

$$d_1 \leqslant a_1 \leqslant d_1' \text{ and}$$
$$d_2 \leqslant a_2 \leqslant d_2' \text{ and}$$
$$\ldots \ldots$$
$$\ldots \ldots \text{ and}$$
$$d_n \leqslant a_n \leqslant d_n'$$

where d_1 and $d_1' \in D_1, \ldots, d_n$ and $d_n' \in D_n$.
If *n*=2, a range search $(d_1 \leqslant a_1 \leqslant d_1')$ and $(d_2 \leqslant a_2 \leqslant d_2')$ corresponds to all points in a two-dimensional space which are inside the rectangle:

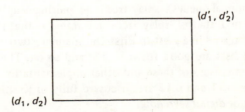

Investigate how the implementation of a PAIR_SET given in this chapter is suitable for range searching?

What types of applications may use range searching?

Are there any other data structures better suited for range searching purposes?

6.19. *Design exercise*

A small technical college wishes to computerise its student records. The system should allow for insertion, deletion and retrieval of student records. It is also required to generate a complete listing of student records in alphabetic order of names, and a list of courses with the names of all the students attending them. Each student record consists of full name, date of birth, term-time and permanent address, unique student number and the courses taken. The college offers not more than 50 courses each year, which are named C101, C102 etc. The expected number of students in the college is approximately 1000. Information regarding students attending courses is subject to continuous change.

Describe an ADT, or a combination of ADTs, which can be used to represent the above data. Discuss the most suitable data structures to implement your ADTs.

Bibliographic Notes and Further Reading

Knuth (1972) is an authoritative encylcopaedia on the subject of hashing. The hashing schemes described here are all static in the sense that the size of the hash table must be pre-determined. Dynamic hashing schemes are being developed where the size of the array can fluctuate (Knuth, 1972; Larson, 1978) (primarily for hash tables on backing memories). More on primary and secondary clusterings together with quadratic quotient hash coding and its effects on clustering can be found in Bell (1970).

Use of priority queues in scheduling programs in a multi-user time-shared system can be found in Lister (1984) and Peterson and Silberschatz (1985). There are many other implementations of priority queues which, although they give $O(\log n)$ average and/or worst-case time complexities, are still superior to heap structures for particular domains. To mention a few techniques — 'leftist tree', 'two list', 'binomial queues', 'splay tree' and 'pairing heap' are among these implementations. For instance: 'splay trees' are *stable* — that is, elements with equal priorities are treated in a First-In-First-Out manner; 'two lists' are good for sizes < 200; linked lists are good for $n < 10$; and so on. The paper by Jones (1986) gives full comparisons of these and other implementations..

Tries of exercises **6.11** and **6.12** are discussed fully in Knuth (1972) and are also known as *radix* or *digital searching*.

Multi-dimensional search trees were first introduced by Bentley (1975). Partial-match and range-searching operations are discussed and analysed in that peper. Also, a survey is given in Bentley and Friedman (1979).

Another specialised set ADT is the 'Component_Set' or Cset which is a set of disjoint sets. Operations merge and find are defined on Csets which respectively creates union of two sets and finds the set containing an element. This and its implementations are given in chapter 7 (section 7.3), as used in the Kruskal's algorithm.

On implementation of sets, it must be noted that for certain applications it may be necessary to duplicate the set elements in order to achieve an efficient implementation. Implementation of set 'frontier' in Dijkstra's algorithm (chapter 7, section 7.4) is an example of this where a priority queue and a bit-vector are used for processing operations insert, delete, delete_min and member. This issue is also touched on in chapter 8 (section 8.3).

Bell, J. R. (1970). 'The quadratic quotient method: a hash code eliminating secondary clustering', *CACM*, Vol. 13, No. 2, February, pp. 107–109.

Bentley, J. L. (1975). 'Multidimensional binary search trees for associative searching', *CACM*, Vol. 18, No. 9, pp. 509–517.

Bentley, J. L. and Friedman, J. H. (1979). 'Data structures for range searching', *ACM Computing Surveys*, Vol. 11, No. 4, December, pp. 397–409.

Fagin, R. *et al.* (1979), 'Extendable hashing: a fast access method for dynamic files', *ACM Transaction on databases*, No. 4.

Jones, D. W. (1986). 'An empirical comparison of priority queues and event-set implementations', *CACM*, Vol. 29, No. 4, April, pp. 300–311.

Knuth, D. E. (1972). *The Art of Computer Programming. Volume 3: Sorting and Searching*, Addison-Wesley, Reading, Massachusetts.

Larson, P. (1978). 'Dynamic hashing', *BIT*, Vol. 18, pp. 184–201.

Lister, A. M. (1984). *Fundamentals of Operating Systems*, 3rd edition, Macmillan, London.

Peterson, J. L. and Silberschatz, A. (1985). *Operating System Concepts*, 2nd edition, Addison-Wesley, Reading, Massachusetts.

7 Non-linear ADTs—Graphs

In chapter 3 we introduced linear data types as a collection of objects where each object, in general, could have one 'next' object and one 'previous' object. Trees were non-linear data types where the above restriction was relaxed and each object could have more than one 'next' object (children of a node) but, at most, only one 'previous' object (parent of a node). We can further generalise tree structures by allowing an object to have more than one 'previous' object. A graph is such an unconstrained structure where each object may have zero, one or many 'next' and 'previous' objects. This generality obviously adds an extra degree of freedom in structuring data to relate to the structure of real-world situations.

7.1 Definitions and Terminology

A graph G is a set of nodes or vertices V and a set of arcs or edges E. This is usually denoted as G = (V, E). The nodes in V are composed of objects which may contain an element of type elemtype. An arc is a pair of vertices (v, w) where v and w are vertices in V. Thus E is a binary relation defined as a subset of V $*$ V (see section 6.4). Figure 7.1 shows a pictorial representation of a graph G1 = (V1, E1) where V1 = {1, 2, 3, 4} and E1 = {(1, 2), (2, 1), (1, 4), (4, 2), (4, 3), (4, 1), (3, 1)}. The nodes of a graph are usually denoted by numbers 1, 2, . . ., n where n = |V|. When an arc $a \equiv (v, w)$ is in E, an arrow from node v to w pictorially represents this arc. Graph G1 can be interpreted as a representation of one-way streets where each node corresponds to a junction and each arc corresponds to a road. A two-way road is represented as two arcs (v, w) and (w, v). Thus, the arcs are *ordered* pairs of vertices – that is, $(v, w) \neq (w, v)$. Such graphs are called *directed graphs* or *digraphs* for short.

In certain situations the pairs need not be ordered; that is, when $(v, w) \equiv (w, v)$. Such graphs are known as *undirected graphs* or simply *graphs* (we use the term 'graph' to refer both to undirected and general graphs. The distinction should be obvious from the context.) Graph G2 = (V2, E2) in figure 7.1 is an undirected graph. V2 = {1, 2, 3, 4} and E2 = {(1, 2), (1, 3), (2, 4), (3, 2), (2, 1), (3, 1), (4, 2), (2, 3)}. However, the last four arcs in E2 are equivalent to the first four arcs. Thus, in the diagrammatical representation of G2 arrows are omitted and arcs are assumed to be two-way. G2 can be interpreted as an instance of a map where nodes are towns and arcs are roads between towns.

Nodes and arcs of a graph may be composed of objects with additional information attached to them. For example, in G1 each arc may have a 'distance' associated with it. Such graphs are called *labelled* or *weighted graphs* (see figure

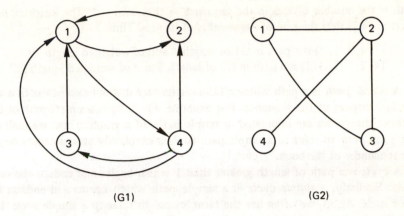

(G1)   (G2)

Figure 7.1 Pictorial representation of graphs G1 and G2.

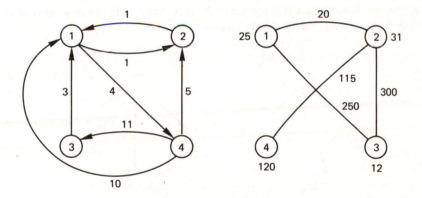

Figure 7.2 Labelled graphs.

7.2). Also, in G2, the population (name etc.) of each town may be attached to that node. When $a \equiv (v, w) \in E$ in a graph G, a is said to be *incident* on v and w, and v and w are said to be *adjacent*. In a digraph G, an arc a exists from v to w, and w is said to be *adjacent* to v.

With regard to graphs G1 and G2, some apt questions might be: (1) Can one get from one junction to another? (2) Is it possible to get from any junction to any other junction? (3) Are there circular or alternative routes? (4) What is the shortest route from one node to another? and so on. To formalise such operations on graphs, let us first introduce a few more terms.

A *path* in a graph G is a sequence of vertices $(v_1, v_2, \ldots, v_m)$ such that $(v_1, v_2), (v_2, v_3), \ldots, (v_{m-1}, v_m)$ are all arcs of the graph G. The *length* of this path is the number of arcs in the sequence — that is, $m - 1$. The *weighted path length* of a path is the sum of the weights of its arcs. Thus

$(1, 4, 3\ 1, 2)$ is a path in G1 of length 4 and of weighted length 19
$(4, 2, 3, 1, 2, 3)$ is a path in G2 of length 5 and of weighted length 985

A *simple path* is a path where all the vertices are distinct except the first and the last vertices of the sequence. For example, $(1, 4, 2)$ is a simple path of G1. Almost always, we are interested in simple paths of a graph. Thus, we shall use the term *path* to refer to a simple path (unless excplicitly stated otherwise) in the remainder of the book.

A *cycle* is a path of length greater than 1 which begins and ends at the same node. Similarly, a *simple cycle* is a simple path which begins and ends at the same node. Again, we often use the term 'cycle' to refer to a simple cycle. For example

$(1, 4, 3, 1)$ is a cycle of length 3 in G1

A graph G is *connected* if there is a path from every node to every other node of G (possibly except itself). For instance

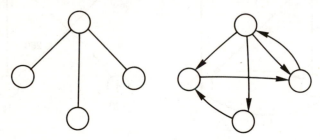

are connected; whereas

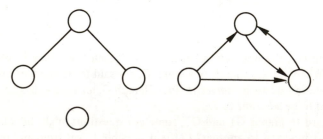

are not connected. If an undirected graph is connected, it will be all in one piece.

A graph G is *acyclic* if it contains no cycles.

A connected acyclic undirected graph is called a *free tree*. This is because it can be transformed into a tree by choosing any of the nodes as a root. In a free tree with n nodes, there must be $n-1$ arcs (see exercise 7.3).

In a graph G = (V, E) a graph of G′ = (V′, E′) is a *sub-graph* of G if V′ ⊂ V and E′ ⊂ E such that if a' ∈ E′ and a' ≡ (v', w') then v' and w' are in V′. If E′ contains all arcs a' ≡ (v', w') such that v', w' ∈ V, then G′ is said to be an *induced* sub-graph of G. Now we can define the ADT graph.

7.2 ADT Graph and its Implementation

As an abstract data type, a graph G = (V, E) consists of a set V, a relation E and a set of operations acting on these two sets. Thus the operations on the ADT graph are precisely those operations of the ADT's set and relation. However, because of the new terminology used, we shall give the set of operations usually used in processing graph data types. In the following, two extra data types — namely, node and arc — are used. A node is an object composed of an element and a label. An arc is an object composed of two nodes and a label. In the following, it is assumed that one (global) instance of the ADT graph is being manipulated. Also, Nelemtype and Nlabeltype refer to the types of elements and the labels of nodes respectively, and Alabeltype refers to the type of labels for arcs.

procedure init_graph;
        set V := { } and E := { }

procedure add_node(n : node);
        set V := V union {n}. Nodes are unique with respect to their
        elements.

procedure add_arc(a : arc);
        set E := E union {a}

procedure set_node_elem(**var** n :node; e : Nelemtype);
procedure set_node_label(**var** n : node; L : Nlabeltype);
        These two procedures initialise a node n to contain an element
        e and a label L.

function label(n : node) : Nlabeltype;
        This function returns the label (weight) of the node n.

procedure set_arc_nodes(**var** a : arc ; n1, n2 : node);
procedure set_arc_label(**var** a : arc ; L : ALabeltype);
        These two procedures initialise the arc a to be from node n1 to
        node n2 with label L. If the graph is undirected, arc a will be
        defined as incident on n1 and n2.

procedure process_nodes(**procedure** P(**var** n : node));
        This prcedure will be used to implement the abstract control
        structure

> **for all** v **in** V **do** P(v);

procedure process_arcs(**procedure** P(**var** a : arc));
> Similarly this is used to implement

> **for all** e **in** E **do** P(e);
> In this book we often use these abstract control structures which can be implemented by introducing a procedure (P) and using these two procedures.

function edge(n1,n2 : node) : boolean;
> If there exists an arc (n1, n2) in E, then a true value will be returned; otherwise a false will be returned.

function weight(n1,n2: node) : Alabeltype;
> If there exists an arc from n1 to n2, function weight returns the label (weight) of that arc. Otherwise a distinguishing value of type Alabeltype is returned.

procedure process_adjacent_nodes(n :node; **procedure** P(**var** x : node;
> **var** a : arc));
> This procedure implements

> **for all** a ≡ (n,x) **in** E **do** P(x,a);

> that is, procedure P(x, a) is applied to all the arcs from node n to other nodes x.

procedure adjacent_nodes(a : arc; **var** n1,n2 : node);
> If a ≡ (v, w), this procedure assigns v and w to n1 and n2 as their values respectively.

Note that, in the above list, not all the operations on sets and relations are mentioned. Other operations such as delete etc. may be added if necessary. However, for our purposes, this list is adequate. Also, note that we have assumed that only one single graph is being processed. If many graphs are being manipulated simultaneously, an extra parameter (**var** G : graph) is needed in all the operations.

Implementation of Graphs

Chapters 5 and 6 give full details of implementing sets and binary relations (PAIR_SETs). To implement V, if it is unbounded, a linked list could be used. For a bounded V an array would be suitable. The 'element' components of the nodes are often positive integers $1, 2, \ldots, n$ where n is the maximum number of nodes in V. Here, a vector of labels can be used to implement *V*.

```
const n =   ;
type node = 1. .n;
    Nlabeltype = . . .; (* Node label type *)
```

var V : **array** [1. .n] **of** Nlabeltype;

To implement relation E in conjunction with the graph operations, there are primarily two approaches which can be used. The choice depends on the frequency of the operations 'edge' and 'process_adjacent_nodes'. Also, storage utilisation is a major deciding factor.

Adjacency Matrix

The arcs are represented as elements of an $n \times n$ matrix M. For an unlabelled graph the element (i, j) of the matrix is a '1' if the arc (i, j) belongs to E. If (i, j) is not in the graph, a '0' would be stored in that location. For labelled graphs, '1's will be replaced by the labels of the arcs and '0's by a distinguishing label (such as ∞). Figure 7.3 shows the adjacency matrices for graphs G1 and G2.

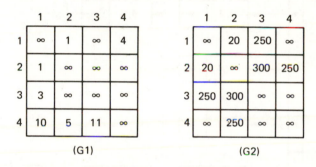

Figure 7.3   Adjacency matrices for G1 and G2.

Note that the matrix for G2 is symmetric. This implies that only $O(n^2/2)$ labels are sufficient to represent an undirected graph with n nodes. An adjacency matrix can be defined as

type matrix = **array** [1. .n, 1. .n] **of** Alabeltype;

In this representation, operations 'edge', 'process_adjacent_nodes' and 'process_arcs' require times of $O(1)$, $O(n)$ and $O(n^2)$ respectively. The storage requirement is $O(n^2)$, irrespective of the size of E.

Adjacency List

To combat the $O(n)$ and $O(n^2)$ time requirements for process_adjacent_nodes and process_arcs, for each node we can store all its adjacent nodes in a list with the corresponding labels of arcs. Using linked lists to implement such adjacency lists, graph G1 may be represented as in figure 7.4.

In adjacency lists, the operation 'edge' requires time proportional to the length of a list. This is the penalty we have to pay for the operation 'process_adjacent_nodes' to execute in a time proportional to the size of the list and not O(n). Also, the storage used is O(e) where e is the number of arcs in E. This suggests that when e is small and thus its adjacency matrix very sparse, this solution should be preferred. However, when e is large (about $n(n-1)/2$) then the matrix solution is superior because extra pointers used for the lists are avoided and the operations are far simpler. For other variations for representing graphs see exercises **7.4, 7.7** and **7.8**. In the next two sections we shall discuss two typical applications of graphs.

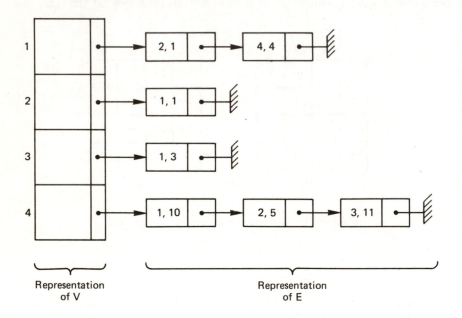

Figure 7.4   Adjacency list representation of G1.

7.3 Spanning Trees

Graph G = (V, E) is a *connected, weighted, undirected* graph. A *spanning tree* (ST) of G is a sub-graph of G, G' = (V, E') such that G' is *acyclic* and connected. Thus a spanning tree of a graph G is a *free tree* which contains all the nodes of G. For instance, in figure 7.5 sub-graphs of G3' and G3" are two spanning trees of graph G3.

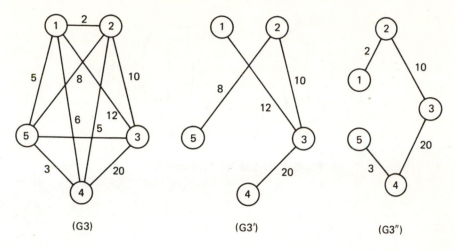

(G3) (G3′) (G3″)

Figure 7.5 Two spanning trees of G3.

The cost of a spanning tree is the sum of the weights of the arcs in that spanning tree. So the cost of G3′ is (10 + 12 + 8 + 20) = 50, whereas the cost of G3″ is (2 + 10 + 20 + 3) = 35. We need to find the Minimum Cost Spanning Tree (MCST) of graph G.

This somewhat abstract problem can have numerous realisations. To mention but a couple:

(a) For constructing a computer network it is required to interconnect *n* computers in such a way that any two computers can communicate, possibly through other computers. Furthermore, we wish to minimise the total cost of construction. The nodes of a graph may denote the computers and the arcs may denote a possible communication link with its associated cost of construction. The MCST of this graph gives the set of links to be constructed for the cheapest network.

(b) A similar graph can be used for designing telephone exchanges for *n* locations.

Before discussing two algorithms for finding the MCST of a graph, let us prove a theorem which will be the basis of our solutions. *Note that* there may be more than one MCST for a given graph.

MCST Theorem

G = (V, E) and U is a proper subset of V, that is, U ≠ V. If the arc (u, v) has the lowest cost among arcs such that $u \in U$ and $v \in V-U$, then there exists an MCST that contains the arc (u, v). (See figure 7.6.)

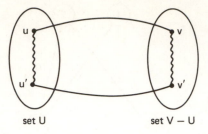

set U   set V − U

Figure 7.6   The MCST theorem.

Proof

We prove the above theorem by contradiction. Let us assume that there exists no MCST containing the arc (u, v). Given an MCST (M, say), we proceed to construct a new spanning tree as follows. Since M is connected, there must be an arc (u', v'), $u' \in U$, $v' \in V-U$. Since M is an ST, there exists a path from u to v; for instance, a path from u to u' followed by the arc (u', v') followed by a path from v' to v. To construct a new ST, M', we remove (u', v') from M and instead we add (u, v) to M. M' will be connected and cannot contain any cycles. So indeed it is a spanning tree. The cost of M' is, however, less than or equal to the cost of M since the weight of (u, v) is less than or equal to that of (u', v') (by definition). Therefore we have constructed a new spanning tree whose cost is less than or equal to that of M'. Since M was an MCST, this implies that M' must also be an MCST. This contradicts our initial assumption. Therefore there exists an MCST containing (u, v).

This theorem can be used in two different ways to build an algorithm to find a practical solution.

Kruskal's Algorithm

The Kruskal's algorithm works by starting with an empty spanning tree (no arcs) and gradually building up an MCST by taking the next lowest arc (in weight) and adding it to the spanning tree, provided that it does not cause a cycle. If it produces a cycle in the ST, it is rejected and the next lowest arc is tried. At each repetition the nodes can be partitioned into U and V − U, and the MCST theorem ensures the correctness of the result (see exercise **7.10**). In G3 the three smallest arcs are (1, 2) with weight 2, (4, 5) with weight 3, and (2, 4) with weight 5. These can be added to the solution as in figure 7.7(a), (b) and (c). The next smallest arc is (1, 5) with weight 5. This arc cannot be added to the solution since it causes a cycle (1, 2, 4, 5, 1). Similarly, the next smallest arcs (1, 4) and (2, 5) cannot be added to the solution. The next arc (2, 3) should be added, which will complete the spanning tree. The cost of this MCST is $(2 + 5 + 3 + 10)$ = 20.

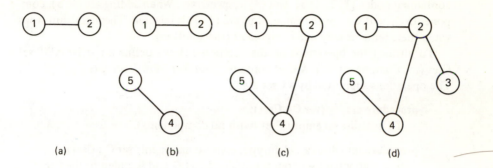

Figure 7.7 The Kruskal algorithm applied to G3.

An outline of Kruskal's algorithm can be defined as

```
var MCST : zet;        (* set of arcs *)
procedure Kruskal;
  var a : arc;
      arcs : priority_queue;
  begin
    MCST := {};        (* initialise MCST as an empty set *)
    arcs := {};        (* initialise arcs as an empty set *)
    for all e in E do insert(e,arcs);
    while MCST is not complete do
        begin
          a := delete_min(arcs);
          if arc a does not cause a cycle in MCST then
              insert(a,MCST)
        end
  end;   (* of Kruskal *)
```

MCST is a set of arcs. Arcs is a priority_queue where arcs can be deleted in order of non-increasing weight. The algorithm works by initialising MCST and arcs and building up a priority queue of all the arcs. Then, repeatedly, the next smallest arc (in weight) is deleted and, if it does not produce a cycle in MCST, it is inserted in MCST.

It now remains to refine the algorithm to stop cycles being formed. To do this, we observe that Kruskal's algorithm initially contains *n components*, each being an MCST with one node only. As the processing continues, components get merged together to form larger components. In each iteration of the loop an arc $a \equiv (v, w)$ is selected. Let v and w be in the components C_v and C_w respectively. Obviously if C_v is the same component as C_w, adding the arc a causes a cycle in the solution. For instance, in figure 7.7c there are two components containing nodes $\{1, 2, 4, 5\}$ and $\{3\}$ respectively. When adding arc $(1, 5)$, components C_1 and C_5 are the same, thus arc $(1, 5)$ is rejected. The algorithm stops when there remains only one component (the solution).

To abstract the operations on the components, we define a special ADT set Cset (for Component set). A Cset contains a number of *disjoint* sets $C_1, C_2, \ldots$. The operations supported by a Cset are

procedure empty (**var** C : Cset);
 To initialise an empty Cset (with no components).

procedure initialise (a : elemtype; com : component; **var** C : Cset);
 The component set com is initialised to $\{a\}$ and is added to the Cset C.

function find (a : elemtype; C : Cset) : component;
 This function returns the component in C which contains element a.

procedure merge (C1, C2 : component; **var** C : Cset);

 This procedure forms C3, the union of the sets C1 and C2, removes C1 and C2 from the Cset and renames C3 as C1 or C2 arbitrarily and adds it to the Cset C.

function size (C : Cset) : integer;
 This function returns the number of components in C.

The type component is used to identify components, and elemtype refers to type of the individual elements.

Using a Cset where the components are identified by numerals $1, 2, \ldots$, etc., the Kruskal algorithm can be refined as

```
procedure Kruskal;
   var a : arc;
       arcs : priority_queue;
       C : Cset;
       x,y : node;
       comx, comy : component;
   begin
       MCST := (); arcs := ();
       empty(C);
       for all e in E do insert(e,arcs);
       (* create a component for each node *)
       for all v in V do initialise(v,v,C);
       (* iterate until C contains only one component *)
       while size(C) > 1 do
         begin
           a := delete_min(arcs);
           adjacent_nodes(a,x,y);
           comx := find(x,C);
```

```
    comy := find(y,C);
    if comx <> comy then
        begin
            insert(a,MCST);
            merge(comx, comy, C);
        end
    end
end; (* of Kruskal *)
```

Notice that this algorithm assumes that the initial graph is connected. Otherwise the **while** loop needs to be modified slightly. Also, the type component is assumed to be compatible with the type of node (0. .*n*) for a bounded graph (see initialise(v, v, C) in above). The mechanics of this algorithm are shown below for the graph G3:

$MCST$	C
{ }	{ {1}, {2}, {3}, {4}, {5} }
{(1, 2)}	{ {1, 2}, {3}, {4}, {5} }
{(1, 2), (4, 5)}	{ {1, 2}, {3}, {4, 5} }
{(1, 2), (4, 5), (2, 4)}	{ {1, 2, 4, 5}, {3} }
{(1, 2), (4, 5), (2, 4), (2, 3)}	{ {1, 2, 3, 4, 5} }

Implementing Csets

A simple way to implement a bounded Cset where the total number of elements is less than or equal to *n* is to use an array of component memberships:

```
type elemtype = 1..n;
     component = 0..n;
     Cset = record
                membership : array[elemtype] of component;
                size : 0..n;
            end;
```

Using this data structure, empty(C) requires a time of $O(n)$ to set all memberships to 0; initialise(a, com, C) requires $O(1)$; find(a, C) requires $O(1)$; and, finally, merge(c1, c2, C) requires $O(n)$ since it has to scan the entire membership array to change any c1 to c2 (or equally any c2 to c1). To perform $n-1$ merge operations, a time of $O(n(n-1)) = O(n^2)$ will be needed. Note that $n-1$ is the maximum number of merge operations which can be performed before there will be only one component left in the Cset.

We can reduce the time needed for the merge operation considerably by structuring elements of each component into a list such that, when merging two components, the entire array elements do not need to be examined. These lists can be established by modifying the Cset structure:

```
Cset = record
        membership : array[1..n] of
                          record
                             com, next : component
                          end;
        components : array[1..n] of component
      end;
```

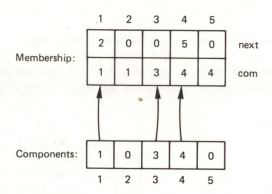

The above diagram shows an instance of a Cset when there are three components — that is, $1 : \{1, 2\}$; $3 : \{3\}$ and $4 : \{4, 5\}$. Now, if components 1 and 3 are to be merged together, component 1 can be merged into component 3 by traversing elements of component 1 (that is, elements 1 and 2) and changing their memberships to 3. Furthermore, component 1 should be removed:

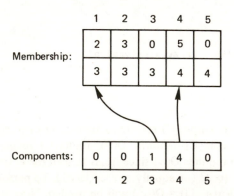

To improve the timing further, it can be observed that when merging component C1 into component C2, all elements of C1 must be traversed. To reduce the overall time requirement we can ensure that the *smaller* component is merged into the larger one on every occasion. To implement this, we merely need to keep the size of each component in the array 'components'. This has little effect on an individual merge operation. But, for maximum of $n-1$ merges, $O(n^2)$ will be reduced to $O(n \log n)$. We observe that the time for $n-1$ merge operations is proportional to the total number of times that memberships of elements are changed. Since smaller components are merged into larger ones, each time that a membership of an element v is changed, v will be in a new component that is at least twice the size of its initial component. Therefore the membership of an element v cannot be changed more than $(\log n)$ times since, after that, there will be only one component left. The total number of membership changes cannot therefore exceed $(n \times \log n)$, for n elements.

Analysis of the Kruskal Algorithm

Returning to the Kruskal algorithm, the **while** loop will be executed at most e times (e is the number of edges in E), since in each iteration one element is deleted from the set of arcs. Choosing a heap data structure for the priority queue *arcs*, the delete_min can be executed in a time of $O(\log n)$. The entire loop therefore needs a maximum time of $O(e \log n) + O(n \log n)$. The latter term is the time needed for all the merge operations. The initialisation of the algorithm needs only $O(n+e)$. The total time needed is therefore $O(e \log n + n \log n + n + e)$ which is equal to $O(e \log n)$. Note that $e \geqslant n-1$ since the graph is a connected graph.

Prim's Algorithm

The MCST theorem can be used in a more direct way to build an algorithm to find the MCST of a graph. Suppose V = {1, 2, 3, . . ., n}. Choose U = {i} for some i : $1 \leqslant i \leqslant n$. Now, the shortest edge (u, v) such that $u \in$ u and $v \in$ V$-$U will be an arc of an MCST for the graph. Set U := U union {v}. Repeat, finding the shortest arc (x, y) such that $x \in$ U, $y \in$ V$-$U and add this to the solution, and finally add node y to the set U. At any instant, set U contains nodes which form an MCST for those nodes. In each iteration one extra arc is added to the final solution. The process terminates when U = V. Since an MCST is a free tree, this iteration will be executed $n-1$ times to reach the final solution. The following shows a diagrammatic representation of the sets U and V$-$U. The arc (x, y) is the shortest (in cost) among all arcs connecting a node in U to another node in V$-$U.

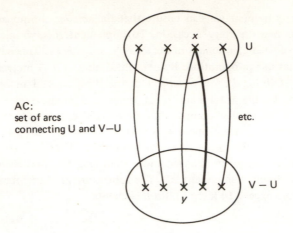

AC:
set of arcs
connecting U and V−U

etc.

Let us denote the set of *arcs* connecting nodes in U to nodes in V − U by AC. The mechanics of Prim's algorithm are illustrated below for the graph G3. We initialise U to contain node 1. As a result, the set AC will contain the four arcs emanating from node 1. The arc (1, 2) has the smallest cost, 2, and it is therefore added to MCST. Now node 2 is added to U and the set AC is reconstructed. The two arcs (1, 5) and (2, 4) have the smallest cost, 5. We can choose either of them. Let us choose the arc (1, 5). Arc (1, 5) is added to MCST and node 5 is inserted in U. The shortest arc in AC is now the arc (5, 4) with cost 3. It is added to MCST and now U becomes {1, 2, 5, 4}. In the new AC, the two arcs (5, 3) and (2, 3) have the smallest cost, 10. Let us choose the arc (5, 3). After adding node 5 to U, U will be equal to V and therefore the process stops with the MCST = {(1, 2), (1, 5), (4, 5), (3, 5)} with cost 20. Note that there is more than one such MCST, and indeed we have a different solution to the one produced by Kruskal's algorithm.

U	AC	MCST
{1}	{(1,2), (1,3), (1,4), (1,5)}	{(1,2)}
{1,2}	{(1,3), (1,4), (1,5), (2,3), (2,4), (2,5)}	{(1,2), (1,5)}
{1,2,5}	{(1,3) (1,4), (2,3), (2,4), (5,3), (5,4)}	{(1,2), (1,5), (5,4)}
{1,2,5,4}	{(1,3), (2,3), (5,3), (4,3)}	{(1,2), (1,3), (5,4),
{1,2,5,4,3} ≡ V		(5,3)}

An outline of Prim's algorithm is given below:

```
procedure Prim;
   var i : integer;  AC : zet;   (* set of arcs *)
   begin
     MCST := {};
     U   := {1};
```

```
AC := {(x,y) | (x,y) is in E and (x=1 or y=1)}
for i := 1 to n-1 do          (* MCST is a free tree *)
    begin
        a := delete_min(AC); (* find the shortest arc in AC *)
        (* a ≡ (u,v) and v ∈ V-U *)
        insert(a,MCST);        (* add a to MCST *)
        insert(v, U);          (* add v to the set U *)
        reconstruct set AC
    end
end;  (* of Prim *)
```

Implementing this algorithm necessitates refining the reconstruction of AC in the **for** loop. AC = $\{(x, y) \mid x \in U$ and $y \in V-U\}$, noting that $(x, y) \equiv (y, x)$ in an undirected graph. An obvious choice would be to use a priority queue to represent AC. Since AC cannot contain more than e arcs, therefore delete_min needs a time of $O(\log e)$. When an arc (u, v) is deleted, restructuring AC will involve removing arcs (v, x) from AC, where $x \in U$, and then adding arcs (v, y) to AC, where $y \in V-U$:

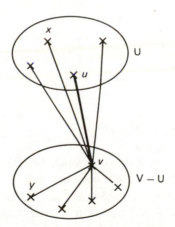

A priority queue, however, does not support such search operations very effectively. An alternative choice for the set AC is a linear list. Again, the above searching actions would necessitate a time of $O(ne)$. In either case, at least $O(ne) \leqslant O(n^3)$ will be necessary.

Prim suggested a data structure for the set AC which reduces the total time complexity to $O(n^2)$, irrespective of the number of arcs e. At each instance, the set AC stores only arcs (i, j) such that for a given $i \in V - U$, the arc (i, j) has the smallest cost among the arcs (i, k) for all $k \in U$:

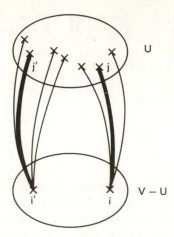

That is to say, only the heavy arcs (shortest arcs emanating from i, i' etc.) are stored in AC. This, in turn, implies that AC cannot have more than n arcs, since $|V-U| \leq n-1$. Therefore we can use an array A to represent the set AC as

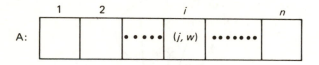

where $A[i] = (j, w)$ when $i \in V - U$

$j \in U$

w is the cost (weight) of arc (i, j)

and j is *closest* node in U to i.

Now, restructuring array A needs only a time of $O(n)$. Also, finding the shortest arc in A takes $O(n)$. To expand the **for** loop of the algorithm, let us assume L is a large value and $L' < L$ is a second large value (not occurring as the weight of an arc). If an arc (i, j) does not exist in the graph, we assign L' to its weight. The body of the **for** loop can be redefined to

```
for i := 1 to n-1 do
    begin
1 ... find A[y] ≡ (x,w) with smallest w in the array
      Add (y,x) to MCST
      Set A[y] := (x,L).  This is to prevent processing the
                          yth element again.
2 ... Update the array to reflect the migration of y from
      V - U to U.  It is only necessary to scan the array
```

```
and check if adding y to U produces a closest arc
to any node in V - U, i.e.

for j := 1 to n do
    (* A[j] ≡ (k,w') *)
    if (weight(y,j) < w') and (w' < L) then
        modify A[j] := (y, weight(y,j))
end;
```

Statements 1 and 2 take $O(n)$ for a linear scan of the entire array. Thus the entire loop takes $O(n(n-1)) = O(n^2)$. The table below shows how the array A is changed when Prim's algorithm is applied to graph G3. Only the elements which are changed are shown. The underlined entries are the selected weights in the MCST.

	1	2	3	4	5	*MCST*
Array A:	(1,L)	(1,2)	(1,12)	(1,6)	(1,5)	(1,2)
		(1,L)	(2,10)	(2,5)	(4,3)	(2,4)
			(2,L)	(2,L)	(4,L)	(5,4)
						(3,2)

Kruskal's algorithm takes $O(e \log n)$. If e is large – that is, $O(n^2)$ – this would be $O(n^2 \log n)$. Therefore for dense graphs, Prim's $O(n^2)$ algorithm is faster than Kruskal's (asymptotically). For small e, however, Kruskal's algorithm is asymptotically superior.

7.4 Path-finding Algorithms

In this section we shall look at a few problems concerning finding paths in digraphs and weighted digraphs. The solutions we shall discuss will equally apply to undirected graphs. In an abstract form we pose four problems which intuitively seem to be in increasing order of complexity:

Problem 1: Find a path from a source node S to a target node T.

Problem 2: Find a shortest path from a source node S to a target node T.

Problem 3: Find a shortest weighted path from a source node S to a target node T.

Problem 4: Find the shortest weighted path from S to all other nodes of the digraph.

As we will show later, the algorithms for the above four problems are structurally similar. So we shall start with problem 4 and derive an algorithm for it which will then be used to derive solutions for the first three problems.

Dijkstra's Algorithm

Weighted Shortest Path length problem (WSP).
Given a weighted digraph $G = (V, E)$ where each arc has a *non-negative* weight, find the shortest weighted path lengths from node S (a source node) to any other node (see later for discussion on obtaining the actual paths).

Dijkstra's algorithm is an incremental solution since it proceeds by constructing a set 'permanent' of nodes x with definite (fixed) shortest weighted path length T_x. Initially, permanent contains node S since its shortest path length is known to be zero ($T_s = 0$). At each stage we consider a set 'frontier' of all the nodes which are not in permanent but are adjacent to a node in permanent. If node y in frontier is adjacent to nodes $x_1, x_2, \ldots$ in permanent, we shall temporarily set T_y to be the minimum of (T_{x_1} + weight of (x_1, y)), (T_{x_2} + weight (x_2, y)), ... Figure 7.8 shows a pictorial representation of the set permanent and the set frontier.

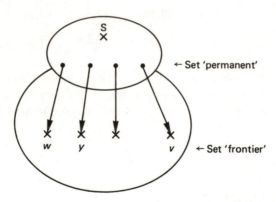

Figure 7.8 Nodes in 'frontier' are directly reachable from 'permanent'.

WSP Theorem

If node w in frontier has the smallest temporary weighted path length T_w over all nodes in frontier, then its temporary weighted path length can be made permanent.
Proof. The proof is by contradiction. Let us assume that T_w cannot be made the permanent shortest weighted path length of node w. Then there must be another path from S to w with a smaller weighted path length. One such path is shown in figure 7.9.

The new path includes the shortest path from S to x (say), then it goes to v, then it follows a sequence of nodes (represented by a curly line) until it reaches

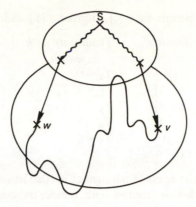

Figure 7.9 Alternative path from S to *w*.

w. In other words T_v + weighted path length of (curly path from *v* to *w*) $< T_w$. But this inequality cannot hold because the weighted path length of the path from *v* to *w* is at least zero (since arc weights are non-negative). Therefore, our initial assumption must have been wrong and indeed T_w is the permanent shortest weighted path length of *w*.

Returning to Dijkstra's algorithm, the WSP theorem implies that node *w* can be moved to the set permanent. A new frontier set can be reconstructed and this process should continue until frontier becomes empty. To fix our ideas, let us look at a concrete example. Let us find the weighted shortest path length from node 1 to nodes 2, 3, 4 and 5 of the digraph G4 of figure 7.10.

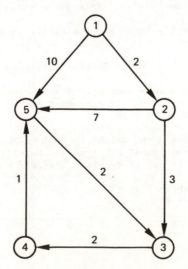

Figure 7.10 Digraph G4.

We begin by initialising permanent = { }, frontier = {1} and $T_1 = 0$.

'permanent'	'frontier'	Temporary path length	Fixed path length
{ }	{1}	$T_1 = 0$	$T_1 = 0$
(i) {1}	{2, 5}	$T_2 = 2, T_5 = 10$	$T_2 = 2$
(ii) {1, 2}	{5, 3}	$T_5 = 9, T_3 = 5$	$T_3 = 5$
(iii) {1, 2, 3}	{5, 4}	$T_4 = 7$	$T_4 = 7$
(iv) {1, 2, 3, 4}	{5}	$T_5 = 8$	$T_5 = 8$
(v) {1, 2, 3, 4, 5}	{ }		

The only node in frontier is made permanent in (i). The nodes adjacent to 1 are 2 and 5 which are added to frontier with temporary costs $T_2 = 0+2 = 2$ and $T_5 = 0+10 = 10$. In (ii), node 2 with the smallest temporary cost is removed from frontier and added to permanent. Nodes 5 and 3 are adjacent to 2, T_5 is changed to the smaller cost T_2 + weight $(2, 5) = 2+7 = 9$, and T_3 is set to $T_2 + 3 = 5$. In (iii), node 3 is moved from frontier to permanent. Node 4 is added to frontier with cost $T_3 + 2 = 7$. In (iv), node 4 with the smaller T_4 is made permanent. Node 5 can now take the smaller temporary cost $T_4 + 1 = 8$. Finally, in (v), node 5 is made permanent. The final shortest weighted path lengths are shown in the last column in the above table.

Now we can give an outline of the Dijkstra's algorithm using the sets frontier and permanent.

```
        procedure Dijkstra;
          var frontier : zet;  x,w : node;
          begin
            for all v in V do T_v := ∞;   (* a large value *)
            T_s := 0;
            (* permanent := {}; This is not strictly necessary *)
            frontier := {S}

(1)...  while frontier <> {} do (* while frontier is not empty do *)
          begin
(2)...        choose and delete node w in frontier with minimal T_w;
              (* insert w in the set permanent *)
(3)...        for all (w,x) in E do
                begin
                  (* modify the weight T_x; modify_weight(x, frontier) *)
(4)...            T_x := minimum(T_x, T_w + weight (w,x));
(5)...            If x is not in frontier then add x to frontier
                end
          end (* of while loop *)
      end; (* of Dijkstra *)
```

Note that the operations on the set permanent are given as comments, since they are not essential to the working of the algorithm.

To finalise this algorithm we need to choose an appropriate set structure for frontier. The **while** loop (1) can be executed at most n times since in each execution of its body, one element is removed from the set frontier. This implies that statement (2) is executed at most n times. The body of the **for** loop (3) processes an arc (w, x) and in the worst case will execute once for every arc of the graph. The statements (4) and (5) are therefore executed at most e times (e is the number of arcs). The operation 'add x to frontier', however, can be executed n times in the worst case, since $|V| = n$. Notice that the analysis of the **for** loop is done independently of the outer **while** loop.

The set frontier must support operations delete_min (2), insert (5), test for membership (5) and update_weight (4). Note that frontier is a set of nodes (x say) which have their corresponding weights (T_x). The delete_min is defined in terms of these weights. The statement (4) in the algorithm requires modifying the weight of a node x (i.e. T_x) to $T_x +$ weight(w, x). It would be advantageous to choose a priority queue to represent the set frontier, but testing for membership and modify_weight can take up to $O(n)$. To combat this problem, one can easily represent the frontier set as a priority queue (primarily to support delete_min and insert) and a set of indices into the priority queue (for each node in the graph) which would aid the modify_weight operations. Here we assume that set V is bounded and is represented as the set $\{1, 2, \ldots, n\}$. With these modifications, the Dijkstra's algorithm can be refined as:

```
procedure Dijkstra;
  const n = ? ;
  type node = 1..n;
  var frontier : zet;
      T : array[node] of integer;
      x,w : node;
  begin
    for all v in V do T[v] := ∞;
    T[S] := 0;
    frontier := {S};
    while frontier <> () do
      begin
        w := delete_min(frontier);
        for all (w,x) in E do
          if T[x] > T[w] + weight(w,x) then
            begin
              T[x] := T[w] + weight(w,x);            .......... (*)
              if not member(x,frontier) then
                  insert(x, frontier)
              else modify_weight(x, frontier)
            end
      end (* of while *)
  end; (* of Dijkstra *)
```

Now we need to design data structures to implement the set frontier. As we indicated earlier, the set frontier can be represented as a heap H and a vector of indices X into the heap. Using this vector, given a node, its location in the heap can be determined in O(1) time.

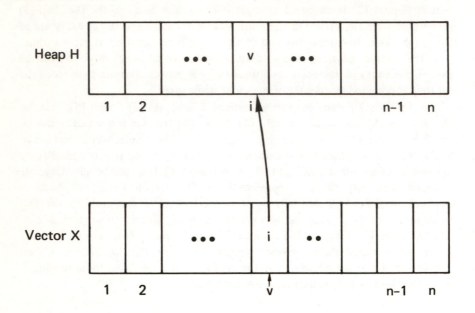

$X[v]$ is zero if node v is not in the heap H. If node v is stored in the location $H[i]$, then $X[v]$ is assigned the value i.

Now, the operations on the set frontier can be implemented as the following:

Insert(x , frontier) :

Use $T[x]$ as the priority of node x and insert x into the heap H. (see section 6.3.2). Each time an element of the array H is updated (H[i] := j), maintain the index vector X accordingly (X[j] := i).

Delete_min(x, frontier) :

Follow the algorithm of section 6.3.2 with the addition that each time a location i in the heap is overwritten by j, maintain the index vector by executing X[j] := i. Also, since x is deleted from the set frontier, its index should be deleted (X[x] := 0).

Member(x, frontier) :

If $X[x]$ is zero, x is not in frontier; otherwise x is a member of frontier.

Modify_weight(x, frontier) :

The new weight for x is given by $T[x]$. Using a similar strategy to insert and delete_min in heaps, element x must be moved towards the root of the partially ordered tree corresponding to the given heap if $T[x] <$ $T[H[X[x]$ div $2]]$. Otherwise, if $T[x] > T[H[X[x]$ div $2]]$ then element x should be successively exchanged with the smallest of its children, i.e. $H[2X[x]]$ and $H[2X[x] + 1]$ and so on. Again, each time an element of the array H is updated ($H[i] := j$), the index array X should be maintained accordingly ($X[j] := i$). (Note that in the Dijkstra's algorithm the latter case will never arise since the temporary path lengths will never increase.)

The worst case time requirement of the above operations are therefore:

insert $\quad$ $O(\log n)$
delete_min $\quad$ $O(\log n)$
member $\quad$ $O(1)$
update_weight $\quad$ $O(\log n)$

According to these data structures the time complexity of Dijkstra's algorithm would be:

$$O(n \quad + \quad n \log n \quad + \quad e \quad + \quad \max(n \log n, \quad e \log n))$$

initialisation $\quad$ delete_min $\quad$ update T_x $\quad\quad$ insert $\quad\quad$ update_weight
$\quad\quad\quad\quad\quad\quad\quad\quad\quad\quad\quad$ and member $\quad$ in frontier

Thus the total worst case time requirement is $O(\max(n, e) \log n)$. Since usually $e > n$, this simplifies to $O(e \log n)$.

So far, this algorithm computes the shortest weighted path lengths for each node. To retrieve the actual paths, we observe that, for each node, it is sufficient to maintain the previous node which led to the shortest weighted path length for that node. We can maintain these in a mapping 'predecessor' for each node. The mapping can be implemented as an array in the above algorithm, and where marked by an asterisk (*) we add 'predecessor[x] := w;'. The path to a node x can then be retrieved by looking up the predecessor nodes:

```
procedure path (x : node);
  begin
    if x <> S then
      begin
        path (predecessor[x]);
        write (predecessor[x])
      end
  end;   (* of path *).
```

A complete Pascal program which would read a graph and find the shortest weighted paths using Dijkstra's algorithm is given in the 'solution to selected exercises' section in chapter 7, no. 20. Of the four problems posed in the beginning of this section we have tackled problem 4. The other three problems can be solved using the same framework. The main differences are that in problems 1 and 2, a much simpler implementation of frontier will be needed and therefore the time complexity will improve considerably (see exercises **7.15** and **7.16**). For problem 3, as soon as the shortest weighted path length for T is found, the loop of Dijkstra's algorithm can be terminated.

In this chapter, we have looked at two common applications of undirected graphs and digraphs. In chapter 10 we shall look at a few more algorithms on graphs.

Exercises

7.1. Give examples of graphs where

(a) *Cyclic* arcs, such as

are needed.

(b) *Parallel* arcs, such as

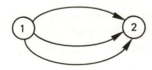

are needed (such graphs are called *coloured* graphs).

7.2. How can a free tree be transformed into a tree structure?

7.3. Prove by induction that a free tree with n nodes has $n-1$ arcs:

(a) True if $n = 1, 2, \ldots$.
(b) If true for $n = x$ then true for $n = x+1$.

7.4. How can symmetric matrices be implemented without wasting any storage? Does your solution affect the time complexity of any graph operation? Distinguish between dense and sparse matrices.

7.5. Give a full implementation of the ADT graph:

(a) using adjacency lists
(b) using adjacency matrices.

7.6. How can parallel and cyclic arcs be represented using adjacency lists and matrices?

7.7. *Incidency matrix*: An $n \times e$ matrix can be used to represent a graph where the element (i, j) is a '1' if node i is incident upon arc j. Define graph operations on incidency matrices. What are the advantages and disadvantages?

7.8. In adjacency lists for undirected labelled graphs, labels will be represented twice in the lists. How can this be avoided? *Hint:* use a multi-list.

7.9. Formulate the following problems in terms of graphs:

(a) To build the cheapest telephone exchange network. The cost of building an exchange between any two cities is known.
(b) On a printed circuit board, many points are to be connected to zero voltage. What is the cheapest way to achieve this, assuming that the points can be interconnected in any way on the board?

7.10. Using the MCST theorem, verify that Kruskal's and Prim's algorithms do work successfully.

7 11. Give full implementation of Csets using arrays.

7.12. Csets can be implemented using a forest of trees where each tree corresponds to a component and the nodes in a tree are the elements of that component. How can such trees be organised to process find and merge operations efficiently?

7.13. Using Kruskal's and Prim's algorithms find the MCST of the following graph:

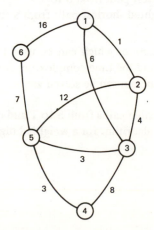

7.14. Using Dijkstra's algorithm find the shortest weighted paths from node 1 to all the other nodes of the digraph

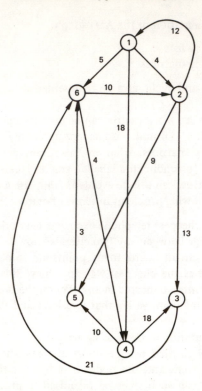

7.15. How can Dijkstra's algorithm be modified to solve the following problems?

(a) To find a path from S to T.
(b) To find the shortest path from S to T.
(c) To find the weighted shortest path from S to T.

7.16. Discuss the data structures which can be used for your solutions to exercise **7.15**, and then find the time complexities for each problem.

7.17. The shortest path problem can be solved as:

Generate all possible paths from S to T and choose the shortest
Design such an algorithm for a weighted digraph. What would its time complexity be?

7.18. *Practical assignment*

A 'useless' production in a context-free grammar is one which cannot be used in the generation of a sentence in the language defined by the grammar. Detection of useless productions is a valuable exercise if only because it might highlight clerical errors in the grammar which would otherwise have gone unnoticed.

Definition. Let G = (N, T, P, S) be a context-free grammar. The production $A \to \alpha$ is said to be *useful* if $S \to * uAw \to u\alpha w \to * uvw$ for some terminal strings *u, v, w*. Otherwise the production $A \to \alpha$ is said to be *useless*.

Example

Let G = ({S, U, V, W}, {a}, P, S), where P consists of

$$S \to aS \quad S \to W \quad S \to U$$
$$W \to aW \quad U \to a$$
$$V \to aa$$

The useful productions in this grammar are

$$S \to aS \quad S \to U \quad U \to a$$

The production $V \to aa$ is useless because there is no way of introducing V into a sentential form (that is, $S \not\to * uVw$ for any *u, w*). The productions $W \to aW$ and $S \to W$ are useless because W cannot be used to generate a terminal string (that is, $W \not\to * w$ for any $w \in T^*$).

The problem of determining which productions are useful splits up into two distinct problems. These are:

Problem 1. For each non-terminal A determine whether $A \to * v$ for some terminal string *v*; that is, determine whether A is a *terminating* non-terminal.

Problem 2. For each non-terminal A, determine whether $S \to * \beta A \gamma$ for some strings β, γ; that is, determine whether A is an *accessible* non-terminal.

A program to remove useless productions will therefore involve computing the terminating non-terminals and removing all productions involving non-terminating non-terminals *followed by* computing the accessible non-terminals and removing all productions involving inaccessible non-terminals.

Hint: We can model the problem as a graph-searching problem in the following way. Given a grammar G, construct a graph H in which the set of nodes is N, the set of non-terminals of G. There is an arc (A, B) in H if there is a production $A \to \alpha B\beta$ where α, β are arbitrary strings. From the definitions of path, and $\Rightarrow *$, $S \Rightarrow * \alpha A\beta$ for some α, β if and only if there is a path from S to A in the graph H.

Figure 7.11 shows the graph so constructed for the grammar with productions:

S → aB   S → BC
A → aA   A → c   A → aDB
B → DB   B → C
C → b   D → B

From the graph we see that S, B, C and D are accessible from S but A is
not. Thus the productions A → aA, A → c and A → aDb are all useless.
Design ADTs for storing and manipulating grammar rules as described.
Hence write a program which would read a set of production rules and
decide which rules are useless.

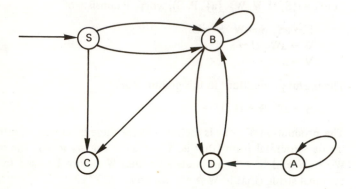

Figure 7.11   Graph showing accessible non-terminals.

7.19. Design an algorithm to find the longest path from a source node to a
destination node in a digraph. What is the time complexity of your algorithm?

Bibliographic Notes and Further Reading

The two MCST algorithms were from Kruskal (1956) and Prim (1957). Yao
(1975) discusses an $O(e \log \log n)$ algorithm for MCST. The Cset implementation
using trees (exercise **7.12**) can be found in Hopcroft and Ullman (1973). The
Dijkstra algorithm originally comes from Dijkstra (1959). This version is a slightly
modified algorithm from Backhouse (1982). Backhouse uses the set 'reached' to
be the union of frontier and permanent. Dijkstra's algorithm does not work
when the costs are negative. Floyd's algorithm (Floyd, 1962) finds the shortest

weighted path lengths from any node to any other node in $O(n^3)$. Floyd's algorithm works if some of the arcs have negative costs, but no cycle with negative costs should be present in the digraph. The solution to practical assignment exercise **7.18** involves a path finder and can be found in Backhouse (1979). To abstract the traversal of all elements of sets V and E and also the set of adjacent nodes to a given node, an alternative method is to realise each set as a list and use operations get_first and get_next. This method together with implementation issues is discussed in Ebert (1987).

Backhouse, R. (1979). *Syntax of Programming Languages: Theory and Practice*, Prentice-Hall, Englewood Cliffs, New Jersey.

Backhouse, R. (1982). 'An analysis of choice in program design', in Biermann and Guiko (eds), *Proceedings of Nato Summer School on Automatic Program Construction*, Reidel, Dordrecht.

Dijkstra, E. W. (1959). 'A note on two problems in connection with graphs', *Numerische Mathematic*, Vol. 1, pp. 269–271.

Ebert, J. (1987). 'A versatile data structure for edge-oriented graph algorithms', *CACM*, Vol. 30, No. 6, pp. 513–519.

Floyd, R. W. (1962). 'Algorithm 97: shortest path', *CACM*, Vol. 5, No. 6, p. 345.

Hopcroft, J. E. and Ullman, J. D. (1973). 'Set merging algorithms', *SIAM Journal of Computing*, Vol. 2, No. 4, pp. 294–303.

Kruskal, J. B. (1956). 'On the shortest spanning subtree of a graph and the travelling salesman problem', *Proceedings of AMS*, Vol. 7, No. 1, pp. 48–50.

Prim, R. C. (1957). 'Shortest connection networks and some generalizations', *Bell System Technical Journal*, Vol. 36, pp. 1389–1401.

Yao, A. C. (1975). 'An $O(|E| \log \log |V|)$ algorithm for finding minimum spanning trees', *Information Processing Letters*, Vol. 4, No. 1, pp. 21–23.

PART II:
ALGORITHM DESIGN
WITH ABSTRACT DATA
TYPES

8 Techniques for Developing Efficient Algorithms

There are no general and standard techniques for designing efficient algorithms for a given problem. However, as you may have realised, we have been using certain useful techniques which have quite often helped us in designing efficient algorithms. For instance, binary search algorithms, tree operations and the quicksort algorithm all have a common theme: they divide the problem into two sub-problems of approximately equal size, and then recursively solve the two sub-problems. The best results are usually achieved when the problem is divided into two equal-sized sub-programs. Such a paradigm is known as *divide and conquer* and is discussed at length below. There are some other techniques which we shall discuss. It must be noted, however, that these techniques are merely guidelines and are not guaranteed to work for every problem. Nevertheless, when solving a new problem it is often useful first to try such a technique in the hope that it will give either a good solution or perhaps some insight regarding the problem properties.

8.1 Divide and Conquer

This technique has been used extensively throughout the book. All tree algorithms are indeed instances of this general approach. The characteristics of divide and conquer algorithms can be summarised as in figure 8.1. The divide and conquer algorithm S for solving a problem P first divides it into k sub-problems $P_1, P_2, \ldots, P_k$ (for a given k). Then each sub-problem is solved. Finally, the problem P is solved by combining the solutions of the sub-problems. This approach when used recursively often gives efficient algorithms for problems in which the sub-problems are smaller versions of the original problem.

Divide and Conquer Approach to Algorithm Design

```
S(P) :
  begin
    (* Divide the problem P into subproblems :
        P₁, P₂, P₃, ..... ,P_{k-1}, P_k (for some k) *)

    (* Solve the subproblems recursively *)
```

```
S(P₁); S(P₂); ..... ; S(Pₖ);

(* Combine the results and hence return the result of P *)
end;
```

Naturally, the size of the sub-problems, P_i, and their number, k, are very crucial when determining the time requirement of the resulting algorithm. Let us look at the binary search algorithm of chapter 1 which determines if x belongs to the sequence $a_1, a_2, a_3, \ldots, a_n$.

```
(* Find x in the sequence a_..a_ *)
search (x, a_L..a_U) :                              time
  begin
    (* If the sequence is not empty, then *)
    if U < L then   failure
    else   (* Compare x and a_(L+U) div 2 *)        [1]
        if x = a_(L+U) div 2 then success
        else if x > a_(L+U) div 2 then
                search(x,a_(L+U) div 2 +1..a_n)    [1+T(n/2)]
            else   (* x < a_(L+U) div 2 *)
                search(a_L..a_(L+U) div 2 -1)      [1+T(n/2)]
  end;
```

The maximum time requirement is

$$T(n) \approx k + T(n/2) \quad \text{for some constant } k$$
$$\approx k + T(n/2)$$
$$\approx kj + T(n/2^j) \quad n/2^j \geqslant 1$$

Since searching stops when the size of the sequence is 1:

$$n/2^j = 1 \Rightarrow j = \log n$$

Therefore

$$T(n) = k \log n + 1$$

This algorithm divides the sequence into two equal-sized sub-sequences. Now, if the algorithm divided the sequence $a_L . . a_U$ into two sequences $a_L . . a_L$ and $a_{L+1} . . a_U$, then it can easily be verified that $T(n) = O(n)$. In fact, this new algorithm corresponds to a sequential search from the beginning of the array. We could try dividing the sequence into α sub-sequences of n/α elements each. Although the base of the logarithm in the above formula increases, its constant coefficient also increases and the total time will exceed the time requirement of the binary algorithm (see exercise 8.1). These examples clearly show that the sizes of the sub-problems should be equal, in other words, the sub-problems should be 'balanced'. This property was overtly emphasised in the tree search problems where the divide and conquer principle was employed. The following examples show further how the divide and conquer technique can reduce the time complexities and also how care must be taken when dividing a problem.

Example 1. MinMax problem: find the minimum and maximum elements of a sequence of n elements $a_1, a_2, \ldots, a_n$

Let us assume that the number of comparisons of the elements is taken as a measure of time requirement. An easy solution involves a linear scan to find the minimum and maximum elements after $2(n-1)$ comparisons. Therefore $2n-2$ comparisons are needed.

The divide and conquer method divides the sequence into two sequences each with $n/2$ elements (assume n is a power of 2). Then the algorithm finds the minimum and maximum of each of these two sequences by recursion, and hence calculates the minimum and maximum of the whole sequence. The listing that follows gives the details of this algorithm.

Divide and Conquer Algorithm for MINMAX

```
procedure MINMAX(S : sequence; var min, max : integer);
   (* This procedure returns min and max which are minimum
      and maximum elements of S respectively *)
   var min1, min2, max1, max2, a, b : integer;
   begin
      if size(S) < 1 then error
      else if size(S) = 1 then
            begin
               (* let S = (a) *)
               min := a; max := a
            end
         else if size(S) = 2 then
               (* let S = (a,b) *)
               if a > b then begin min := b; max := a end
               else begin min := a; max := b end
            else
               begin
                  (* Divide the sequence S into two sequences
                     S1 and S2 each with n/2 elements *)
                  MINMAX(S1, min1, max1);
                  MINMAX(S2, min2, max2);
                  if min1 > min2 then begin min := min2
                  else min := min1;
                  if max1 > max2 then begin max := max1
                  else max := max2
               end
   end;
```

$T(n)$ the number of comparisons can be calculated:

$$T(n) = \begin{cases} 1 & n = 2 \\ 2T(n/2) + 2 & n > 2 \end{cases}$$

Two sub-problems for every problem are constructed:

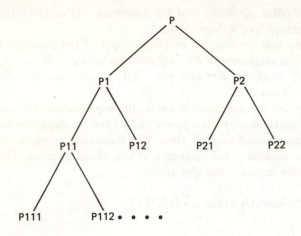

$$T(n) = 2(2T(n/4) + 2) + 2$$
$$= 2^2 T(n/4) + 2^2 + 2$$
$$\cdot$$
$$\cdot$$
$$\cdot$$
$$= 2^k T(n/(n/2)) + 2^k + 2^{k-1} + \ldots + 2 \quad \text{where} \quad 2^k = n/2$$
$$= n/2 + 2^{k+1} - 2 = n/2 + n - 2 = 3n/2 - 2$$

Therefore the divide and conquer method has reduced the number of comparisons by a constant term to $\underline{3n/2 - 2}$.

Example 2. Multiplication of numbers problem: multiply two *n*-bit integers X and Y

(This can easily be generalised to *n* bytes or *n* words etc.) The simple school method takes $O(n^2)$-bit operations. The divide and conquer method divides the two integers, X and Y, into two *n*/2-bit integers.

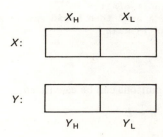

$$X = X_L + 2^{n/2} \times X_H$$
$$Y = Y_L + 2^{n/2} \times Y_H$$
$$X \times Y = (X_L + 2^{n/2} X_H)(Y_L + 2^{n/2} Y_H) = X_L Y_L + 2^{n/2}(X_L Y_H + X_H Y_L)$$
$$+ 2^n X_H Y_H$$

Now we can use a recursive procedure to compute $X \times Y$. There are four sub-problems of size $n/2$ and therefore the time complexity $T(n)$ is given by

$$T(n) = 4T(n/2) + kn \Rightarrow T(n) = O(n^2)$$

This does not lead to any improvement. However, an alternative way of combining the sub-solutions will reduce the time requirement. We can reduce the number of multiplications to three by rewriting the recursive formula as

$$X \times Y = X_L Y_L + 2^n X_H Y_H + 2^{n/2}[(X_H - X_L)(Y_L - Y_H) + X_H Y_H + X_L Y_L]$$

Now a recursive algorithm based on this new formulation of multiplication would give the time complexity $T'(n)$ as

$$T'(n) = \begin{cases} 3T'(n/2) + k'n & n > 1 \quad \text{for some } k' \\ 1 & n = 1 \end{cases}$$

By induction we can show that

$$T'(n) = 3kn^{\log 3} - 2kn$$
$$\Rightarrow T'(n) = O(n^{\log 3})$$
$$= O(n^{1.59})$$

By careful division of the problem into three sub-problems, we have reduced the asymptotic complexity of the algorithm from $O(n^2)$ to $O(n^{1.59})$ (figure 8.1).

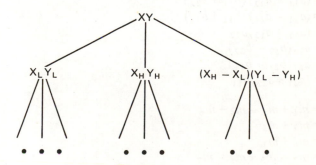

Figure 8.1 The three sub-problems of the multiplication problem.

Our final example is for multiplying matrices.

Example 3. Matrix multiplication problem: (Strassen matrix multiplication algorithm) multiply two $n \times n$ matrices

The conventional method for multiplying two $n \times n$ matrices, A and B, requires $O(n^3)$ addition operations.

Let us now apply the divide and conquer method.

Divide each matrix into 4 $n/2 \times n/2$ matrices:

$$\begin{pmatrix} a_{11} & a_{12} \\ a_{21} & a_{22} \end{pmatrix} \begin{pmatrix} b_{11} & b_{12} \\ b_{21} & b_{22} \end{pmatrix} = \begin{pmatrix} c_{11} & c_{12} \\ c_{21} & c_{22} \end{pmatrix}$$

Where a_{ij}, b_{ij} and c_{ij} are $n/2 \times n/2$ matrices. Let us assume that n is a power of 2. It can be verified that

$$c_{11} = a_{11} \times b_{11} + a_{12} \times b_{21}$$
$$c_{12} = a_{11} \times b_{12} + a_{12} \times b_{22}$$
$$c_{21} = a_{21} \times b_{11} + a_{22} \times b_{21}$$
$$c_{22} = a_{21} \times b_{12} + a_{22} \times b_{22}$$

Thus, computing a c_{ij} requires four additions and eight multiplications of matrices of size $n/2 \times n/2$, and therefore the time complexity for a recursive program would be

$$T(n) = 8\ T(n/2) + 4\ (n/2)^2, \text{ that is, } T(n) = O(n^3)$$

Strassen's method uses the following equations to reduce the eight matrix multiplications of the above formula to seven:

$$m_1 = (a_{12} - a_{22})(b_{21} + b_{22})$$
$$m_2 = (a_{11} + a_{22})(b_{11} + b_{22})$$
$$m_3 = (a_{11} - a_{21})(b_{11} + b_{12})$$
$$m_4 = (a_{11} + a_{12})b_{22}$$
$$m_5 = a_{11}(b_{12} - b_{22})$$
$$m_6 = a_{22}(b_{21} - b_{11})$$
$$m_7 = (a_{21} + a_{22})b_{11}$$

Then $c_{11} = m_1 + m_2 - m_4 + m_6$
$$c_{12} = m_4 + m_5$$
$$c_{21} = m_6 + m_7$$
$$c_{22} = m_2 - m_3 + m_5 - m_7$$

Now we need only 7 multiplications and 18 additions/subtractions of matrices of size $n/2 \times n/2$. Therefore the new time complexity is

$$T'(n) = 7\ T'(n/2) + 18\ (n/2)^2 \quad n \geqslant 2$$
$$T'(n) = O(n^{\log 7}) = O(n^{2.81})$$

Again, the asymptotic time complexity is reduced from $O(n^3)$ to $O(n^{281})$ (see the formulae in chapter 1, section 1.5). If n is not a power of 2, we can embed the matrix into a bigger matrix by adding zero elements in appropriate places.

The examples shown here all involve a one-dimensional search space. The multi-dimensional search trees are examples of a multi-dimensional divide and conquer method. The divide and conquer paradigm can be used in a variety of other problems, some of which are to be found as exercises at the end of this chapter.

8.2 Dynamic Programming

Dynamic programming was initially devised by Bellman (1957) as a technique for evaluating multi-stage processes and decision-making processes. Basically, the slogan of dynamic programming is

"Any sub-policy of an optimum policy must itself be optimum"

In terms of divide and conquer terminology, this would read: "any solution to a sub-problem for an optimum solution to a problem must itself be optimum."

For instance, in the shortest path problem for digraph, if a path (S, a, b, c, T) is optimal, then all sub-paths (S, a), (S, a, b), (S, a, b, c), (a, b, c) etc. must be optimal. Since, if for instance the path (a, b, c) is not optimal, it can be replaced with a shorter path and hence results in a shorter path length from S to T. Indeed, as we shall find out later, Dijkstra's algorithm is an instance of the dynamic programming approach. However, in order to illustrate the concepts we shall concentrate on some simpler examples first.

Problem 1. Order of matrix multiplication problem
We require to evaluate the following sequence of matrix multiplications:

$$M = M_1 \times M_2 \times \ldots \times M_n$$

where M_i is a matrix of order $r_{i-1} \times r_i$. In what order should the product be computed to minimise the total computation required? (We assume that multiplication of an $m \times n$ matrix by an $n \times p$ matrix requires $m \times n \times p$ operations.) Notice that matrix multiplication is associative — that is, $M(M' \times M'') = (M \times M')M''$.

In the following example, there are four matrices to be multiplied:

$$M = M_1 \times M_2 \times M_3 \times M_4$$

M_1: 5×15
M_2: 15×30
M_3: 30×2
M_4: 2×50

That is

$$(r_0, r_1, r_2, r_3, r_4) = (5, 15, 30, 2, 50)$$

Let us try a couple of different orderings:

$$((M_1 \times M_2) \times (M_3)) \times M_4 - \text{requires 3050 operations}$$

$M_1 \times M_2$ needs $5 \times 15 \times 30$ operations and results in M', a 5×30 matrix. Then $M' \times M_3$ needs $5 \times 30 \times 2$ operations and results in M'', a 5×2 matrix. Finally, $M'' \times M_4$ needs $5 \times 2 \times 50$. The total number of operations is therefore $2250 + 300 + 500 = 3050$.

Similarly

$$M_1 \times (M_2 \times (M_3 \times M_4)) - \text{requires 29 250 operations}$$

The problem is therefore to find the order for which the required number of operations is minimal.

A divide and conquer solution can be formulated as follows:

(1) Divide the matrices into two sequences:

$$(M_1 \times M_2 \times \ldots \times M_i) \times (M_{i+1} \times \ldots \times M_n) \text{ for } 0 < i < n$$

(2) Find the optimal orders for the two sub-problems (sub-sequences).
(3) Hence, the two optimal solutions can be combined to give the final solution. This involves all other different groupings of n matrices being considered — see below.

Here again, for the solution to be optimal, the two sub-problems must be optimal. A recursive solution according to the above scheme would lead to an exponential algorithm.

In divide and conquer, a problem of size n is usually divided into sub-problems such that the sum of the sizes of the sub-problems is very small, typically $O(n)$. In these situations a recursive algorithm would give a polynomial time complexity. However, in certain cases (like the matrix evaluation), the obvious division of a problem of size n produces $O(n)$ sub-problems of sizes $O(n)$. In these cases, a recursive solution will lead to exponential time complexities. However, in many such problems, the total number of *distinct* sub-problems is not very large. Figure 8.2 shows the diagram of the procedure calls in such problems, which is a Directed Acyclic Graph (DAG) — see chapter 10 (section 10.4). Each node represents a sub-problem and a directed arc denotes the dependencies among the sub-problems. The number of arcs in this DAG corresponds to the total number of procedure calls in a recursive program whereas the number of nodes corresponds to the total number of sub-problems to be solved. In this matrix problem, there are an exponential number of arcs whereas the number of nodes is a polynomial function of n. Therefore, unlike divide and conquer, in dynamic programming, we adopt a *bottom-up* approach. We solve all the sub-problems from leaves towards the root of the DAG and store them in a table. So, they need not be computed more than once.

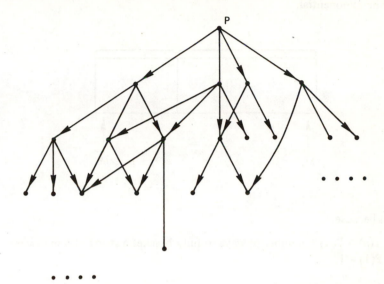

A → B: b is a sub-problem of A

Figure 8.2 DAG of sub-problems in dynamic programming approach.

Generally speaking, for optimisation problems, where we usually try to maximise or minimise a function subject to certain constraints, the dynamic programming approach is:

(1) Consider the problem as many sub-problems. If the problem is to make many decisions simultaneously, the approach is to formulate the problem as a sequential process where some of the decisions have already been made, and then the best decision should be made under those conditions.

(2) When all the sub-problems are solved, as when all partial optimum decisions are known, we can combine these partial solutions to get the optimal solution of the whole problem. The relationship of the sub-problem to the whole problem can usually be defined recursively by a function equation.

For the matrix problem above we must decide how to bracket all n matrices. According to (1), we first decide how to bracket n matrices into two sequences of i and $n-i$ matrices. Then, for each sequence, we make decisions sequentially. Let us, for the time being, concentrate on the minimum number of operations which are needed to compute the product.

If a recursive algorithm is designed based on the divide and conquer formulation of the problem, then the time complexity $T(n)$ (total number of operations) would be exponential.

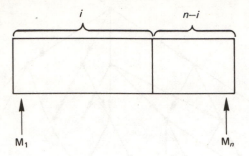

This is because:

$T(n) \propto P(n) \equiv$ number of ways to fully bracket a string of n matrices

$P(1) = 1$

$$P(n) = \sum_{i=1}^{n-1} P(i)P(n-i)$$

This sequence is known as Catalan Numbers and it can be shown that

$$P(n+1) = 1/(n+1) \binom{2n}{n}$$

$$\geqslant 2^{n-1}$$

This shows that the total number of sub-problems (not distinct) to be solved is exponential.

Dynamic Programming Solution to the Matrix Sequence Problem

A bottom-up evaluation of the DAG of sub-problems can be done by solving the sub-problems in order of increasing length of the sub-sequences. Let M_{ij} be the minimum cost of computing $M_i \times M_{i+1} \times \ldots \times M_j$. Then, for i and j ($1 \leqslant i \leqslant n$ and $1 \leqslant j \leqslant n$):

$$M_{ij} = \begin{cases} 0 & i=j \\ \\ \underset{i \leqslant k < j}{\text{MIN}} (M_{ik} + M_{k+1,j} + r_{i-1}r_kr_j) \end{cases}$$

The general case is $(M_i \times \ldots \times M_k)(M_{k+1} \times \ldots \times M_j)$

Cost of sub-problems $\qquad M_{ik} \qquad\qquad M_{k+1,j}$

Dimensions of the

resulting matrices $\qquad (r_{i-1} \times r_k) \qquad (r_k \times r_j)$

The cost of multiplying these two matrices corresponding to the two sub-sequences is $r_{i-1}r_kr_j$ and hence the total cost is $M_{ik} + M_{k+1,j} + r_{i-1}r_kr_j$.

To solve our original problem, we need to compute M_{1n}. Figure 8.3 shows a portion of the DAG of the sub-problems (a few identical nodes are underlined).

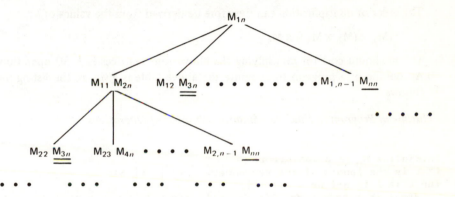

Figure 8.3   DAG for the sequence of matrix multiplications.

Now, we compute the M_{ij} values in order of increasing difference of subscripts — that is, $(j-i)$. Therefore first we compute:

$$
\begin{array}{lll}
 & M_{i,i} & 1 \leqslant i \leqslant n \\
\text{then} & M_{i,i+1} & 1 \leqslant i \leqslant n-1 \\
\;\bullet & \quad \bullet\;\bullet & \quad \bullet\;\bullet\;\bullet \\
\;\bullet & \quad \bullet\;\bullet & \quad \bullet\;\bullet\;\bullet \\
\text{then} & M_{i,i+L} & 1 \leqslant i \leqslant n-L \\
\;\bullet & \quad \bullet\;\bullet & \quad \bullet\;\bullet\;\bullet \\
\;\bullet & \quad \bullet\;\bullet & \quad \bullet\;\bullet\;\bullet \\
\text{and finally} & M_{i,i+n-1} & i = 1
\end{array}
$$

This ensures that when computing M_{ij}, M_{ik} and $M_{k+1,j}$ are computed already. During the course of computation, the solutions to the sub-problems are therefore stored in a table for later use (table 8.1).

Table 8.1 Table of solutions to all sub-problems

$M_{11} = 0$	$M_{22} = 0$	$M_{33} = 0$	$M_{44} = 0$
$M_{12} = 2250$ $k = 1$	$M_{23} = 900$ $k = 2$	$M_{34} = 3000$ $k = 3$	
$M_{13} = 1050$ $k = 1$	$M_{24} = 2400$ $k = 3$		
$M_{14} = 1550$ $k = 3$			

The order of multiplication can therefore be derived from the values of k:

$$(M_1 \times (M_2 \times M_3)) \times M_4$$

The minimum cost for multiplying the above four matrices is 1550 operations. An outline of a program to compute the above table is given in the listing that follows.

Outline of Program to Find the Minimum Number of Operations

```
Initialise M₁ⱼ := 0 for all i;
(* L is the length of the sequences, i.e. |j-i| *)
for L := 1 to n-1 do
    for i := 1 to n-L do    (* for all i compute Mᵢⱼ such that j-i = L *)
        begin
            j := i+L;
            Mᵢⱼ :=   MIN (Mᵢₖ + Mₖ₊₁ⱼ + rᵢ₋₁rₖrⱼ)
                     i ≤ k < j
        end;
write(M₁ₙ)
```

The worst-case time complexity is therefore $O(n^3)$. This is because the inner loop is executed $O(n^2)$ times and the MIN operation in the worst case needs $O(n)$. So the time requirement is reduced from $O(2^n)$ to $O(n^3)$. The space complexity is $O(n^2)$ for the table of solutions.

Problem 2. Knapsack problem

Items $I_1, I_2, \ldots, I_m$ with sizes $S_1, S_2, \ldots, S_m$ are to be packed in a knapsack of capacity C (S_i values and C are positive integers). Pack the knapsack with a collection of these items such that the unused capacity is minimised.

A simple algorithm for this problem would be to generate all possible combinations of the m items (using divide and conquer) and then to select the one with the maximum total size subject to the capacity constraint C. This would obviously be an $O(2^m)$ algorithm (see exercise 8.3).

A dynamic programming solution can be developed by dividing the problem into two sub-problems as follows. Let M_{ij} be the maximum capacity obtained by selecting items from the list $I_1, I_2, \ldots, I_i$ subject to the total capacity being less than or equal to j.

Now we require the value of M_{mc}. In calculating M_{ij} the sequence $I_1, I_2, \ldots, I_i$ can be broken into two sub-sequences $I_1, I_2, \ldots, I_{i-1}$ and I_i. It can be argued that in order to obtain M_{ij}, either I_i is included in the solution or it is not. If it is not included, then $M_{ij} = M_{i-1,j}$. If it is included, before we added the ith item, there must have been enough room for it and therefore $M_{ij} = M_{i-1,j-S_i} + S_i$. Therefore to maximise the capacity:

$$M_{ij} = \begin{cases} 0 & i < 1 \text{ or } j \leqslant 0 \\ M_{i-1,j} & j < S_i \text{ and } i > 0 \\ \text{MAX } [M_{i-1,j}, (M_{i-1,j-S_i} + S_i)] & \text{otherwise} \end{cases}$$

Thus a program to calculate M_{ij} is divided into two sub-programs to calculate $M_{i-1,j}$ and $M_{i-1,j-S_i}$, after which the results are combined. As a recursive procedure, this formula would generate a DAG of procedure calls. Since the optimality of M_{ij} is implied by the optimality of its sub-problems, a dynamic programming algorithm can be used to compute M_{ij} in a bottom-up fashion and store the results in a table. Since $0 \leqslant i \leqslant m$ and $0 \leqslant j \leqslant C$, a $(m+1) \times (C+1)$ table, T, is required. The elements of the table should be computed in the order of increasing values of j and then increasing values of i. Thus

```
Knapsack :
    for i := 0 to m do T₁,₀ := 0;
    for j := 0 to C do T₀,ⱼ := 0;
    for i := 1 to m do
        for j := 1 to C do
            begin
                Tᵢⱼ := Tᵢ₋₁,ⱼ;
                if j >= Sᵢ then
                    if Tᵢ₋₁,ⱼ₋ₛᵢ + Sᵢ > Tᵢ₋₁,ⱼ then
                        Tᵢⱼ := Tᵢ₋₁,ⱼ₋ₛᵢ + Sᵢ
            end;
```

This algorithm is clearly $O(mC)$ as contrasted with the simple $O(2^m)$ algorithm. To fix our ideas, let us look at an instance of the knapsack problem where $C = 8$:

Item	I_1	I_2	I_3	I_4
Size	7	4	6	1

The table T of computation would then be

i \ j	0	1	2	3	4	5	6	7	8
0	0	0	0	0	0	0	0	0	0
1	0	0	0	0	0	0	0	7	7
2	0	0	0	0	4	4	4	7	7
3	0	0	0	0	4	4	6	7	7
4	0	1	1	1	4	5	6	7	8

This table only gives the maximum capacity which can be obtained. A flag can be stored with each element which indicates whether $M_{i-1,j}$ or $M_{i-1,j-s_i}$ was used in computing M_{ij}. From these flags it can then be determined which items led to the maximum capacity (see exercises **8.18** and **8.19**).

Problem 3. Shortest path problem

In chapter 7 we discussed the efficient Dijkstra's algorithm for finding the shortest weighted path length from a source node to a destination node d in a digraph G=(V,E). A recursive algorithm can be formulated by using the dynamic programming approach. T_d, the shortest path length of node d can be expressed as

$$T_d = \underset{(x, d) \in E}{\text{MIN}} \; [T_x + \text{weight}(x, d)]$$

This can be used as a basis of a recursive procedure to compute t_d. However, it would give an $O(n!)$ algorithm where n is the number of nodes in G. Noting that the optimality principle of dynamic programming holds in this situation, a bottom-up evaluation of the DAG of sub-problems leads to a Dijkstra algorithm which would run in $O(e \log n)$. It must be emphasised here that the optimality principle must hold so that the sub-problems can be solved independently of one another. For instance, if we redefine the definition of a path length as the sum of the weights of the arcs modulo W (for a constant W), then dynamic programming could not be used (see exercise **8.22**). The exercises contain a few more problems for which the dynamic programming approach yields efficient algorithms.

8.3 Other Techniques

We have so far looked at two very important programming paradigms. These are general techniques applicable to a variety of problems and data types. However, there are some other techniques which help in the design of efficient algorithms for particular data types or problems.

For solving many graph problems, two traversal techniques (DFS and BFS) systematically visit all the nodes and arcs of a graph and, often efficient algorithms based on these traversals can be developed. Chapter 10 contains many examples of these techniques. Similarly, the tree traversals of chapter 4 can be used as a framework for designing systematic search algorithms on the data type trees.

Another method is the use of 'redundancy' in the data structures. Often, one representation of an abstract data type does not lead to the efficient execution of all the operations. In these circumstances, some or all parts of the data structure can be duplicated in a different structure in order to speed up all the operations. One example is the B^+-trees discussed in chapter 4. There we had a set with both random access and sequential operations. So, the normal B-tree structure was augmented with an extra linear list to provide fast sequential access. Another example was the set frontier in the Dijkstra's algorithm (section 7.4) with the operations Initialise, Insert, Delete_min, Member and Modify_weight. It was implemented as a heap and a vector (or a hash table if necessary).

Lastly, randomised algorithms should be mentioned. Primarily, these algorithms use the ideas of hashing and random functions. The implementation of sets in chapter 6 and the string-matching algorithm of chapter 11 (section 11.3) are instances of this category.

Other techniques usually deal with 'hard' combinatorial problems. These techniques, which include 'exhaustive search', 'branch and bound', 'optimisation' and 'heuristics', are discussed in chapter 12 where a very brief exposition of 'hard' and difficult problems is also included.

Exercises

8.1. Write a program (3-ary search) similar to binary search, but divide the sorted sequence into three sub-sequences. What is the time complexity of your solution? Compare it with the binary search algorithm. Generalise this result for *k*-ary search.

For certain difficult problems, the divide and conquer technique helps in the development of correct algorithms, although the sub-problems may not be of equal lengths. Here are some examples.

8.2. Write a program to generate all permutations of numbers 1. .*n*.
Hint: to get all permutations of numbers 1. .*n*, insert number 1 in all possible positions in the set of permutations of numbers 2 . .*n*.

8.3. The array S is defined as

```
const n = ?;
var S : array [1..n] of boolean;
```

Write a Pascal procedure to generate all combinations of *n* boolean variables stored in the array S.

Hint: To generate all combinations of *n* variables, first generate all combinations of *n*−1 variables, then add to this a false and true value for the *n*th variable.

8.4. Write a recursive function which, given a logical expression with *n* boolean variables and a set of truth values for these variables, would determine whether its value is true or false. Assume that the expression is stored as a binary tree of nodes where

```
type ptr      = ↑node;
     tagtype = (variable, Lnot, Lbinop);
     OpType  = (Land, Lor, Limp, Lequiv);
     node    = record
                 case tag : tagtype of
                 variable : (index : 1..n);
                 Lnot     : (branch : ptr);
                 Lbinop   : (left, right : ptr;
                               binop : OpType)
               end;
```

The variables are indexed into an array [1. .*n*] of boolean and the values therefore can be read immediately. For example

$$p \text{ and } q \equiv \text{not } ((\text{not } p) \text{ or } (\text{not } q))$$

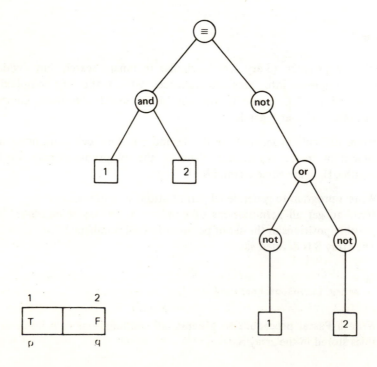

8.5. Using the structures of exercises **8.3** and **8.4**, write a program to decide whether a logical expression is *satisfiable* — that is, if there exists a combination of n boolean values assigned to its variables such that it would yield a true value. What is the maximum time complexity of your algorithm.

8.6. Modify the solution to exercise **8.4** so that it decides whether the logical expression is a 'tautology' — that is, if it is true for all possible combinations of its n variables.

8.7. A binary switch is defined as follows:

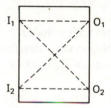

The output can be either (I_1, I_2) or (I_2, I_1) — that is, any permutation of (I_1, I_2). Using this as a building block, it is required to build an 'n-ary' switch, where the output can be any permutation of the n input signals.

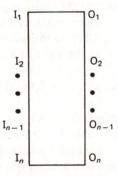

Using divide and conquer, design an n-ary switch. How many binary switches are necessary? (Assume that n is a power of 2.)

Hint: to build an n-ary switch, combine 2 $n/2$-ary switches and $O(n)$ binary switches.

8.8. In a two-dimensional space $[1..n] \times [1..n]$, there are M points. A point P is said to *dominate* another point Q if the first and second co-ordinates of point P are greater or equal to those of point Q. Write a program to compute, for each point, the number of points that it dominates.

Hint: use multi-dimensional divide and conquer. To solve a problem in

k-space (k-dimensional space) of size n, solve two sub-problems of size $n/2$ in k-space, then one problem of size n in $(k-1)$-space.

8.9. *Dynamic programming*

Modify and augment the program to find the minimum number of operations, given under section 8.2, so that it prints out the ordering of the matrices in a format such as:

$$((M_1 \times (M_2 \times M_3)) \times M_4)$$

8.10. Show that $P(n) \geqslant 2P(n-1)$ and therefore $P(n) \geqslant 2^n$.

8.11. Show that $P(n+1) = \dfrac{1}{n+1}\dbinom{2n}{n}$ and hence show that $p(n) \geqslant 2^{n-1}$.

8.12. Show that there are $P(n)$
  - different binary trees with n nodes.
  - ways to divide a polygon into triangles.

8.13. What is the space complexity of Dijkstra's algorithm as a dynamic programming algorithm?

8.14. A problem similar to the matrix multiplication is Optimal Weighted binary search Trees (OWT). Earlier in the book (chapter 4) we discussed Huffman's algorithm to construct a tree which optimises the external weighted path length. In OWT, keys $k_1 < k_2 < k_3 < \ldots < k_n$ are given with weights $w_1, w_2, \ldots, w_n$. It is required to construct a *binary search tree* which minimises $\sum_i w_i L_i$ where L_i is the path length of node k_i.

Hint: the solution is similar to the solution given above for the sequence of matrix multiplication in that a node k_r can be chosen as the root and $k_i \ldots k_{r-1}$ should form an OWT to the left of this node and nodes $k_{r+1} \ldots k_j$ should form an OWT to the right of k_r. This results in a very similar recursive formula.

8.15. Maximise $[f_1(x_1) + f_2(x_2) + \ldots + f_n(x_n)]$ subject to $x_1 + x_2 + \ldots + x_n = M$

where x_i are positive integers ($i=1 \ldots n$) and $f_i(0) = 0$ and all functions are monotonically increasing — that is, $x > y$ implies that $f_i(x) > f_i(y)$.

Assume that the values for each function are stored in an array.

An instance of this problem is when f_i is the interest function of bank i and one wishes to maximise the interest when investing M amount of money.

8.16. *Minimum sum segment problem*

A sequence of integers $I_1, I_2, \ldots, I_n$ is given. A *segment* is defined as a sub-sequence $I_k, I_{k+1}, \ldots, I_{k+j}$ ($1 \leqslant k \leqslant n$ and $1 \leqslant k+j \leqslant n$). Write an algorithm which finds such a segment whose sum of its elements

$$\sum_{k \leqslant x \leqslant k+j} I_x$$

is minimal for some k and j.

8.17. In the knapsack problem, if only the maximum attainable capacity is required, what would the space and time complexity of the algorithm be?

8.18. In the knapsack problem, modify and augment the algorithm to produce the set of items selected in the solution.

8.19. *Weighted knapsack problem*

A knapsack with capacity C is to be filled with items with given capacities $n_1, n_2, \ldots, n_k$, and values $v_1, v_2, \ldots, v_n$, such that the total value is maximised subject to a total capacity of C.

Write an algorithm for finding the set of items maximising the total value. What is the time complexity of this solution?

8.20. Calculate $\binom{n}{m}$ using the equality:

$$\binom{n}{m} = \binom{n-1}{m} + \binom{n-1}{m-1}$$

The dynamic programming approach results in a Pascal triangle. Compare the time complexity of this algorithm with an algorithm based on

$$\binom{n}{m} = n! \, / \, m! \, (n{-}m)!$$

8.21. Show that a recursive shortest path algorithm needs a time of $O(n!)$ in the worst case.

8.22. For the following digraph, find the shortest path lengths from node 1 to all other nodes where path lengths are defined as the sum of the arc weights modulo 7.

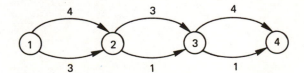

Bibliographic Notes and Further Reading

The original concepts of dynamic programming come from Bellman (1957). The book by Hu (1982) has extra material on dynamic programming.

Dynamic programming can be thought of as an instance of 'tabulation' techniques, where certain values are stored in a table while developing the solution. The string-searching algorithms of chapter 11 are based on tabulation techniques. The recursive sequence of chapter 1 and the Fibonacci series are examples of this. It is also worth while to note that in functional programming languages where a function is recursive and its top-down evaluation produces a DAG of function evaluations, tabulation and bottom-up techniques can be used to get effective solutions. The application of dynamic programming in the evaluation of recursive functions in functional programming contexts is discussed in Bird (1980).

Multi-dimensional divide and conquer is further advocated by Bentley (1980). The problem in exercise **8.7**, and many others, are discussed there. Catalan Numbers are further discussed in Gardner (1976).

Bellman, R. E. (1957). *Dynamic Programming*, Princeton University Press, California.

Bentley, J. (1980). 'Multi-dimensional divide and conquer', *CACM*, Vol. 23, No. 4, pp. 214–229.

Bird, R. S. (1980). 'Tabulation techniques for recursive programs', *Computing Survey*, Vol. 12, No. 4, December, pp. 403–417.

Gardner, M. (1976). 'Catalan Numbers', *Scientific American*, June, pp. 120–124.

Hu, T. C. (1982). *Combinatorial algorithms*, Addison-Wesley, Reading, Massachusetts.

9 Sorting: An Algorithm on the ADT List

Sorting is an old and well-known problem with well-studied and analysed solutions. Here we shall use it as an example of algorithm design and discuss the choices which can be made regarding its implementation.

The ADT list is a linear data type of elements of a given type (elemtype). A list L is given, consisting of n elements:

$$a_1, a_2, a_3, \ldots, a_{n-1}, a_n$$

A relation '$\leqslant$' (less than or equal to) is defined on the elements of type elemtype such that, for a pair of elements a and b, either $a \leqslant b$ or $b \leqslant a$ (or both). The problem of sorting is defined as rearranging the elements of the list L as

$$a_{i_1}, a_{i_2}, a_{i_3}, \ldots, a_{i_{n-1}}, a_{i_n}$$

such that

$$a_{i_1} \leqslant a_{i_2} \leqslant a_{i_3} \leqslant \ldots \leqslant a_{i_n}$$

where $i_1, i_2, i_3, \ldots, i_n$ represents a permutation of numbers $1, 2, 3, \ldots, n-1, n$. For instance, if the elements are names of people, a sort would arrange a list of names into alphabetical order if '$\leqslant$' is defined as the usual alphabetical ordering of names. If the elements are composite objects (persons) each consisting of a name and a salary and the relation '$\leqslant$' is defined to be that of salaries, a sort would give a list of people in order of non-decreasing salaries.

For our discussion, we shall assume that the elements to be sorted are of type integers and furthermore that the lists are implemented as arrays of elements. Thus

```
const n =  ;    (* The number of elements in the list *)
type elements : array[1..n] of integer;
```

will apply throughout this chapter. In section 9.4 we shall investigate a specialised algorithm for sorting lists which are implemented as linked lists. Most of the algorithms are equally applicable to lists of real numbers or character strings.

9.1 An Abstract Sorting Algorithm

We use the divide and conquer approach to algorithm design to develop an abstract sorting algorithm. Given a list L to be sorted, first we 'split' L into two

257

sub-lists, L1 and L2. These two lists are then sorted as two smaller lists, and finally they will be 'joined' together to make a complete sorted list. A recursive formulation of this abstract algorithm can be written in Pascal:

```
procedure sort (var L : list);
  var L1, L2 : list;
  begin
    if size(L) > 1 then
      begin
        split(L, L1, L2);
        sort(L1);
        sort(L2);
        join(L1, L2, L)
      end
  end;
```

This procedure is doubly recursive. It takes a list L and returns it as a sorted list. The terminating condition is that the size of the list should be less than or equal to one. The function size(L) returns the number of elements in the list L. The procedure split(L, L1, L2) takes a list L and 'splits' it into two lists L1 and L2. The procedure join(L1, L2, L) 'joins' the two lists L1 and L2 to make list L.

Figure 9.1 shows a diagrammatic representation of this abstract sort algorithm.

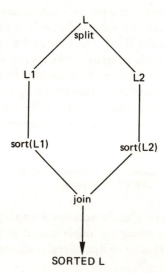

Figure 9.1   Divide and conquer algorithm sort.

The definitions of procedures 'join' and 'split' are specified with no reference to how they may be implemented. This is because, depending on exactly how these operations are defined, we get different sorting algorithms with distinct characteristics. Let us look at an example:

List L: 4   12   3   9   1   21   5   2

List L has 8 elements. There can be two distinct ways of splitting this list into two lists L1 and L2. The first divides the list in the middle to get two lists:

L1:   4   12   3   9
L2:   1   21   5   2

Thus the splitting is based on a *position* in the list. The recursive part of the sort algorithm will sort these two lists to give

L1:   3   4   9   12
L2:   1   2   5   21

Now, we have two sorted lists and 'join' simply consists of a 'merge' operation ('union' of two disjoint sets – see chapter 5 on ordered lists). Therefore, merging these lists results in

L: 1   2   3   4   5   9   12   21

The second solution is more subtle, splitting being based on the *content* or values of the elements of the list. For instance, we can split the list L such that all the elements in L1 are smaller than those in L2:

L1:   4   2   3   1
L2:   9   21   5   12

Following this approach, these lists are sorted recursively to give

L1:   1   2   3   4
L2:   5   9   12   21

Now, 'joining' simply consists of *concatenating* lists L1 and L2 to give list L:

sorted L: 1   2   3   4   5   9   12   21

In the former approach, splitting was easy and the major work was done in the joining process. By contrast, in the latter approach joining was a relatively simple operation and the bulk of the work was carried out while the list was being split. We shall call these approaches *Easy Split/Hard Join* (ES/HJ) and *Hard Split/Easy Join* (HS/EJ) algorithms respectively. In ES/HJ, splitting is based on *position*, whereas in HS/EJ splitting it is based on *content*. Each approach can further be refined to give a variety of sorting algorithms, each with particular features and properties.

9.2   Easy Split/Hard Join Sorting Algorithms

This class of sorting algorithms splits a list L into two sub-lists using a position p in the list as a pivot point. Thus, L1 contains all elements before position p up to and including the element at position p, and L2 contains all elements after position p. The order of elements within each list is preserved as in the original list.

   We further classify these algorithms depending on which position is used as a pivot point to split the list. The two methods are: (1) choose p to be the middle of the list such that lists L1 and L2 are of equal size; (2) choose p to be 'previous (last(L))' such that list L1 will have $n-1$ elements and list L2 will have only 1 element. These lead to two algorithms known as merge sort and insertion sort respectively.

Merge Sort

Let us assume that the elements of list L are stored in array A and that the lower and upper indexes of the array define the list L.

```
type list = record (* lower and upper indices in the array A *)
              lower, upper : 1..n
            end;
var  A : elements;
     L : list;          (* List to be sorted *)
     temp : elements;   (* A temporary list used for
                           joining two sorted lists, see later *)
```

Thus, L.lower = 1 and L.upper = n. The operations size and split can now be given:

```
function size (L : list) : integer;
  begin
    size := L.upper - L.lower + 1
  end;   (* of size *)

procedure split (L : list; var L1, L2 : list);
  begin
    with L do
      begin
        L1.lower := lower;
        L1.upper := (upper + lower) div 2
        L2.lower := L1.upper + 1
        L2.upper := upper
      end
  end;   (* of split *)
```

So, each time a list of size n is divided into a list of the first (n **div** 2) elements, and a list of the remaining $n-(n$ **div** 2) elements. This operation needs a time of O(1), irrespective of the value of n.

The 'join' operation, however, is more complex. In this context, every time join(L1, L2, L) is called, the two lists L1 and L2 are consecutive in the array A. We indicated in section 9.1 that this operation is primarily a 'merge' operation on two ordered lists L1 and L2. Therefore, using a single scan of the two lists we can form the 'join' of them (see intersection(S1, S2, S) of chapter 5).

However, as we scan the two lists, the resulting list should be stored in a temporary array (temp) which should then be copied back into A.

```
procedure join (L1,L2 : list; var L : list);
  var I : integer;
  begin
    (* merge elements A[L1.lower] .. A[L1.upper] and
       elements A[L2.lower] .. A[L2.upper] and store
       the result in temp[L1.lower] .. temp[L2.upper] *);
    for I := L1.lower to L2.upper do    (* copy temp to A *)
      A[I] := temp[I]
  end;  (* of join *)
```

This procedure needs a time of O(m) where m is the size of lists L1 and L2. Figure 9.2 shows the mechanics of the merge sort. Each curly bracket corresponds to a 'join' and the underlined elements are the list operands of that 'join' (for the details of the merge operation above, see exercise **9.2**).

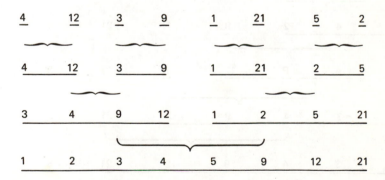

Figure 9.2 Operation of the merge sort.

The time complexity of the merge sort can be derived using the recurrence formula:

$$T(n) = 2T(n/2) + O(n)$$

where $T(n)$ is the time for sorting a list of n elements. $2T(n/2)$ is for sorting two sub-lists and $O(n)$ is for joining the two sub-lists. From the formulae of section 1.5, it can be shown that $T(n) = O(n \log n)$. This is in fact a very good algorithm. However, it requires the extra array 'temp', which means that its space requirement is $O(n)$ (an extra $O(\log n)$ space also being required for the stack in recursion). Therefore, for large lists we seek solutions without using any excess storage. It should be noted that the merge sort requires the same amount of time and space in the best case, the worst case and the average case.

Insertion Sort

In insertion sort, a list is split into a *singleton list* and a list with $n-1$ elements. Indeed, we can apply the same operations as we used for merge sort with the exception that 'split' returns a list containing the first $n-1$ elements and a list containing the last element. Figure 9.3 shows the working of this algorithm as the 'join' procedures are called.

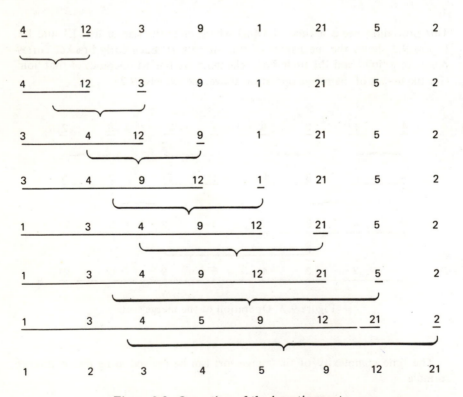

Figure 9.3 Operation of the insertion sort.

The worst-case analysis of this algorithm can be derived:

$$T_{max}(n) = T_{max}(n-1) + \quad O(1) \quad + \quad O(n)$$

sorting L1   sorting L2   joining L1
and L2

Thus

$$T_{max}(n) = O(n^2)$$

To achieve a lower bound for the best-case analysis, we can simplify the solution by transforming the recursive solution into an iterative one.

As can be seen from figure 9.3, the non-recursive mechanism of the insertion sort is to take one element repeatedly and *insert* it in its proper place in the sorted list of elements before it. For instance, 12 is taken and *inserted* in the list (4) to give (4, 12); 3 is taken and *inserted* in the list (4, 12) to give (3, 4, 12) etc. This is why this algorithm is called insertion sort. Therefore a non-recursive insertion sort can be expressed as

```
for I := 2 to n do
    Insert A[I] in its proper place in
    the sorted sequence A[1] .. A[I-1]
```

Each iteration of this loop is known as a *pass* or *phase*. In pass (I–1), the insertion of an element E at A[I] can be done by comparing it with elements A[I–1], A[I–2], . . in turn and shifting elements one to the right until A[J] is found such that E < A[J], in which case E is dropped in A[J]. For example, in pass 6, element 5 is compared with 21, 12 and 9 and each element is shifted one place to the right until 5 is found to be less than 9 and inserted in its proper place:

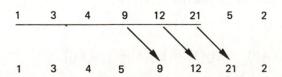

The new non-recursive formulation of the insertion sort is given below:

```
for i := 2 to n do
    begin                       (* pass i-1 *)
    j := i-1;
    E := A[i];                  (* the ith element of the list *)
    while (A[j] > E) and (j > 1) do
```

```
        begin
          A[j+1] := A[j];  (* shift to the right *)
          j := j-1
        end;
   if (j=1) then
      if (A[j] > E) then
         begin
           A[j+1] := A[j];
           j := j-1;
         end;
   A[j+1] := E             (* drop E in its proper place *)
end;
```

In the **while** loop the condition $(j > 1)$ ensures that a reference to $A[0]$ is never generated. The **if** statement checks this special case (see exercise 9.5).

This new solution removes the $O(n)$ storage which was needed for recursion and for the array 'temp'. In the old solution, the best-case and average-case time was always $O(n^2)$. In this non-recursive solution, we will now show that the best-case time reduces to $O(n)$. As a measure of time requirements, we often measure the number of *comparisons* and *data transfers* performed on the elements of the array. In the above solution the number of comparisons C is the number of times $(A[j] > E)$ is executed. C is obviously dependent on the number of times the **while** loop is executed. This is proportional to the number of elements bigger than the ith element to the left of the ith element.

$$C = \sum_{i=2}^{n} (1 + d_i)$$

where d_i is the number of elements bigger than $A[i]$ in the list $A[1] \ldots A[i-1]$.

An extra comparison is needed to stop the loop. Thus

$$C = (n-1) + \Sigma d_i$$

Similarly, the number of data transfers X is

$$X = (n-1) \quad + \quad (n-1) \quad + \sum_{i=2}^{n} d_i$$

$$E := A[i] \quad A[j+1] := E \quad A[j+1] := A[j]$$

$$X = 2(n-1) + \Sigma d_i$$

The best case occurs when $d_i = 0$ — that is, when the list is already sorted. The worst case happens when $d_i = i-1$ — that is, when the sequence is in reverse order.

$$C_{min} = n-1 \qquad\qquad X_{min} = 2(n-1)$$

$$C_{max} = \frac{n^2 + n - 2}{2} \qquad X_{max} = \frac{n^2 + 3n - 4}{2}$$

For the average case, we have to assume that all $n!$ permutations of n elements are equally likely. Thus, on average, $d_i = (i-1)/2$ and therefore

$$C_{avg} = \frac{n^2 + 3n - 4}{4} \qquad X_{avg} = \frac{n^2 + 7n - 8}{4}$$

To summarise, the insertion sort needs $O(n)$ in the best case and $O(n^2)$ in the worst and average cases. Therefore this algorithm is biased towards *nearly sorted* lists and must be preferred when such a bias is known to exist in the original list.

9.3 Hard Split/Easy Join Sorting Algorithms

As discussed in section 9.1, this class of algorithms splits a list L into two lists, L1 and L2, based on the contents of elements such that all elements in L1 have smaller values than those in L2. As the example of section 9.1 showed, the main work is done in splitting the list and joining the sub-lists is a somewhat trivial job of concatenating two lists.

Thus, a list of n elements

L: $a_1, a_2, a_3, \ldots, a_n$

is divided into two lists:

L1: $a_{i_1}, a_{i_2}, \ldots, a_{i_m}$

and

L2: $a_{i_{m+1}}, a_{i_{m+2}}, \ldots, a_{i_n}$

where $a_{i_j} < a_{i_k}$ if $1 \leqslant j \leqslant m$ and $m+1 \leqslant k \leqslant n$ and $(i_1, i_2, \ldots, i_n)$ is a permutation of numbers 1 to n.

A similar classification as for the ES/HJ can be applied to this class. The splitting algorithm either divides list L into two equal-sized sub-lists (approximately) or divides it into a singleton list and a list of $n-1$ elements. The latter approach leads to selection sort and heapsort and the former leads to quicksort.

Quicksort

Quicksort aims to split a list into equal-sized sub-lists (on average, see later). If list L is stored in an array A (as in the data specification of mergesort), then it can be arranged that the lists L1 and L2 be stored in locations A[1] .. A[m] and A[m+1] .. A[n] respectively. This way the 'join' operation is virtually trivial. The quicksort algorithm can therefore be simplified as

```
quicksort (L) : if size(L) > 1 then
                begin
                  split(L, L1, L2);
                  quicksort(L1);
                  quicksort(L2)
                end;
```

The splitting is carried out by a systematic scan of the list from both ends. First, one particular element of the list, say x, is chosen to be the *pivot value*. The splitting is to be done such that all elements of L1 are *smaller* than x and all elements of L2 are *greater than or equal to x*. A simple strategy to achieve this is

(1) Scan the list from the left until an element a_i is found such that $a_i \geqslant x$.
(2) Scan the list from the right until an element a_j is found such that $a_j < x$.
(3) Exchange the elements a_i and a_j.
(4) Repeat (1) (from position $i+1$) and (2) (from position $j-1$) until these two scans meet in the middle of the list.

Thus, the list can be split at position m (somewhere in the middle of the list) such that all elements from position $m+1$ to n are greater than or equal to x. For instance, in our example of L

$$L: \quad 4 \quad 12 \quad 3 \quad 9 \quad 1 \quad 21 \quad 5 \quad 2$$

let us take 5 as the pivot value. A scan from the left finds $12 \geqslant 5$ and a scan from the right finds $2 < 5$.

$$4 \quad 12 \quad 3 \quad 9 \quad 1 \quad 21 \quad 5 \quad 2$$
$$\quad \uparrow \qquad\qquad\qquad\qquad\qquad \uparrow$$
$$\quad i \qquad\qquad\qquad\qquad\qquad\quad j$$

These two elements are exchanged:

$$4 \quad 2 \quad 3 \quad 9 \quad 1 \quad 21 \quad 5 \quad 12$$
$$\quad \uparrow \qquad\qquad\qquad\qquad\qquad \uparrow$$
$$\quad i \qquad\qquad\qquad\qquad\qquad\quad j$$

The scans are repeated and this time, from the left, element 9 is found to be greater than 5 and, from the right, element 1 is found to be less than 5.

$$4 \quad 2 \quad 3 \quad 9 \quad 1 \quad 21 \quad 5 \quad 12$$
$$\qquad\qquad\quad \uparrow \quad \uparrow$$
$$\qquad\qquad\quad i \quad j$$

Exchanging these two elements results in

$$4 \quad 2 \quad 3 \quad 1 \quad 9 \quad 21 \quad 5 \quad 12$$
$$\qquad\qquad\quad \uparrow \quad \uparrow$$
$$\qquad\qquad\quad i \quad j$$

If we repeat our scanning from both ends we get:

$$4 \quad 2 \quad 3 \quad 1 \quad 9 \quad 21 \quad 5 \quad 12$$

$$\uparrow \quad \uparrow$$

$$j \quad i$$

This is where the two scans meet in the middle of the list, indicating that the splitting is complete. All the elements at positions 1 to j are smaller than 5, and elements at positions i (which must be equal to $j+1$) to n are greater or equal to 5.

This splitting operation can be written in Pascal as

```pascal
procedure split (L : list; var L1, L2 :list);
  var i,j : integer;
  begin
     (* choose as pivot value, an element of list L, say x *)
     i := L.lower;         (* initialise the scans from left *)
     j := L.upper;         (* and right of the list *)
     repeat
        while A[i] < x do i := i + 1;  (* A[i] ≥ x *)
        while A[j] >= x do j := j - 1;  (* A[j] < x *)
        if i < j then
           begin
              swap(i,j);  (* to exchange A[i] and A[j] *)
              i := i + 1
              j := j - 1
           end
     until i > j;
     (* list L1 is A[L.lower] .. A[j] and
        list L2 is A[j+1] .. A[L.upper] *)
     L1.lower := L.lower;
     L1.upper := j;
     L2.lower := j + 1;
     L2.upper := L.upper
  end;  (* of split *)
```

The procedure 'swap' exchanges the two elements A[i] and A[j] as

```pascal
procedure swap(i,j : integer);
   var temp : integer;
      begin
         temp := A[i];
         A[i] := A[j];
         A[j] := temp
      end;  (* of swap *)
```

The **if** statement in the procedure split is used to halt exchange of elements when the scans have met in the middle of the list.

To sort our example list completely, we must sort the sub-lists:

4   2   3   1   9   21   5   21

L1   L2

Let us choose the pivot values 3 and 12 for L1 and L2 respectively. The split operation applied to these lists results in

1   2   3   4   9   5   21   12

L11   L12   L21   L22

It now remains to sort sub-lists L11, L12, L21 and L22. Let us take values of 2, 4, 9 and 21 as the pivot values for these sub-lists respectively. The splitting produces eight sub-lists, each with one element.

1   2   3   4   5   9   12   21

L111   L112   L121   L122   L211   L212   L221   L222

Thus the entire list is sorted as

1   2   3   4   5   9   12   21

In this example the procedure split produced two equal-sized sub-lists every time. This may not always be the case. Indeed, this depends largely on the method used to choose a pivot value. In our example, the maximum of the last two unequal elements was chosen as the pivot value. This strategy implies that, for the correct operation of the sorting algorithm, the procedure 'split' should return a status (of type (Ok, NotOk)) value of NotOk when all the elements of the array are equal. In this case, the sort algorithm should be modified to

```
quicksort(L) : if size(L) > 1 then
                  begin
                    split(L, L1, L2);
                    if status = Ok then
                       begin
                          quicksort(L1);
                          quicksort(L2)
                       end;
                  end;
```

It is assumed that variable 'status' is a local variable of quicksort. Note that, in principle, we could have chosen any element, but certain pivot values may not work with our algorithm — see exercise **9.10**.

To find the space and time complexities, let us first look at the operation split. In that procedure a complete linear scan of the list is done (the left and right scans meet in the middle). Thus the time requirement to split a list of size n is $O(n)$.

The best-case time complexity is achieved always when a list is split into two equal-sized sub-lists. This occurs when the pivot value is the *median* of the list. Therefore, list L is divided into two lists of size $n/2$; each of these is divided into two lists of size $n/4$, and so on. So, a tree of procedure calls using the size of the lists can be drawn:

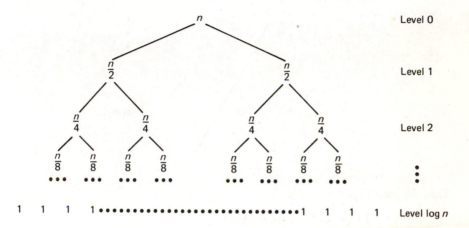

The amount of work done can thus be calculated (for splitting lists):

Level 0: $1 \times n$ $= n$ One list of size n

Level 1: $2 \times \dfrac{n}{2}$ $= n$ Two lists of size $n/2$

Level 2: $4 \times \dfrac{n}{4}$ $= n$ Four lists of size $n/4$

Level $\log n$: $n \times \dfrac{n}{2^{\log n}}$ $= n$ $2^{\log n}$ lists of size 1

Thus the total time needed is $O(n \log n)$

The worst case occurs always when a pivot value is chosen which splits the list into a singleton and a list of the rest of the elements (when the minimum or maximum element is chosen). With our scheme this occurs if the list is already sorted! The tree of procedure calls is now

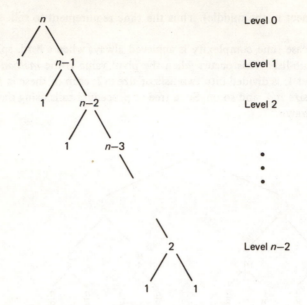

Thus the total time required is

$$n + (n-1) + (n-2) + \ldots + 2 = O(n^2)$$

(Therefore, in the worst case, quicksort is in fact a slow sort!)

Despite this worst-case behaviour, the average-case time complexity turns out to be $O(n \log n)$. To show this, let us assume that the list does not have any duplicate elements. Also we assume that the choice of the pivot value is random, thus each element has a $1/n$ chance of being picked as the pivot value. For each pivot value chosen, the sizes of the sub-lists will be S and $n - S$ (S depends on the pivot value). Obviously, for different pivot values, S can take values from 1 to $n - 1$. Thus the average time complexity is

$$T(n) = \underbrace{n}_{\substack{\text{for split} \\ \text{operation}}} + \underbrace{\frac{1}{n} \sum_{S=1}^{n-1} (\quad T(S) \quad + T(n-S))}_{\substack{\text{average of all possible choices of a} \\ \text{pivot value (for sorting two sub-lists)}}}$$

$$T(n) = n + \frac{2}{n} \sum_{S=1}^{n-1} T(S) \tag{9.1}$$

since

$$\sum_{S=1}^{n-1} T(S) = \sum_{S=1}^{n-1} T(n-S) = (T(1) + T(2) + T(3) + \ldots + T(n-1))$$

Equation (9.1) can be used to show $T(n) = O(n \log n)$ by induction. It is true that for $n = 2$, $T(n) \leqslant n \log n$ (that is, $T(n) = O(n \log n)$). We prove that if $T(S) \leqslant \alpha (S \log S)$ for all $S < n$, then $T(n) \leqslant \alpha (n \log n)$ for all n.

$$T(n) \leqslant n + \frac{2\alpha}{n} \sum_{S=1}^{n-1} S \log S$$

$$\leqslant n + \frac{2\alpha}{n} \left(\sum_{S=1}^{n/2} S \log S + \sum_{S=n/2+1}^{n-1} S \log S \right)$$

Since $\log S \leqslant \log \dfrac{n}{2}$ for $1 \leqslant S \leqslant \dfrac{n}{2}$

and $\log S \leqslant \log n$ for $\dfrac{n}{2} + 1 \leqslant S \leqslant n$:

$$T(n) \leqslant n + \frac{2\alpha}{n} \left(\sum_{S=1}^{n/2} S \log \frac{n}{2} + \sum_{S=n/2+1}^{n-1} S \log n \right)$$

$$\leqslant n + \frac{2\alpha}{n} \log n \left(\sum_{S=1}^{n/2} S + \sum_{S=n/2+1}^{n-1} S \right)$$

$$\leqslant n + \frac{2\alpha}{n} \log n \left(\sum_{S=1}^{n-1} S \right)$$

$$\leqslant n + \frac{2\alpha}{n} \log n \times \frac{n(n-1)}{2}$$

$$\leqslant n + \alpha (n-1) \log n$$

This immediately implies that $T(n) = O(n \log n)$. Indeed, in practice it is not always true that the selection of the pivot value is a random process. However, empirical results show that in these situations the average case is only worse than $(n \log n)$ by a small factor.

From the above analysis it is evident that quicksort is not very good for a list which is nearly in sorted order. In such cases a simpler algorithm such as insertion is more appropriate.

The average complexity of quicksort can be improved firstly by a better choice of pivot values and secondly by observing that for small lists, a simpler sort algorithm is faster than quicksort where the nested loops and recursion times dominate the time for comparisons and exchanges. With regard to the former, ideally the median of the list should be chosen each time. An approximation would be to choose the median of a small sample of elements. A reasonable compromise is to choose the median of the first, last and middle elements of the list.

Abstract Data Types and Algorithms

Following the latter point, for sizes of lists less than or equal to m (a thresh-hold), insertion sort runs faster than quicksort. Therefore, in quicksort, the statement 'if size (L) $>$ 1 **then**' should be changed to

```
if size(L) <= m then
    insertion_sort(L)
else  .........
      .........
```

The actual value for m is dependent on the machine and compiler being used. Experiments show that $5 \leqslant m \leqslant 15$. The actual value can be found after a few trials.

Selection Sort

In HS/EJ sorting algorithms, when the 'split' operation divides the list into a singleton list L1 and a list L2 of the rest of the elements, the resulting algorithm is known as selection sort. Since the splitting is based on the values of elements, the only element of L1 must be less than all the elements in L2 — that is, it must be the minimum element of the list (the only exception being when all the elements are equal). In the recursive sort procedure, the sort(L1) statement is therefore redundant. Also the 'join' operation is no longer needed. Thus selection sort can be presented as

```
selection_sort :
    if size(L) > 1 then
        begin
          split(L, L1, L2);
          selection_sort(L2)
        end;
```

Procedure split ensures that L1 contains the minimum element of L and L2 contains the rest of the elements of L. The tail recursion can easily be trans-formed into a loop structure. The non-recursive solution can now be expressed as

```
selection_sort :
    for i := 1 to n-1 do
        begin
          (* Find the minimum element of A[i] .. A[n] and
             replace it with A[i] *)
        end;
```

For a simple selection algorithm, finding the minimum element can be done by a single scan of the elements $A[i] \ldots A[n]$:

```
S := i;
min := A[S];
for j := i+1 to n do
    if A[j] < min then
        begin
            S := j;
            min := A[S]
        end;
(* Now A[S] is the minimum element of A[i]..A[n] *)
```

The operation of this simple selection algorithm for our example would be:

pass 1	4	12	3	9	1	21	5	2
pass 2	1	12	3	9	4	21	5	2
pass 3	1	2	3	9	4	21	5	12
pass 4	1	2	3	9	4	21	5	12
pass 5	1	2	3	4	9	21	5	12
pass 6	1	2	3	4	5	21	9	12
pass 7	1	2	3	4	5	9	12	21
pass 8	1	2	3	4	5	9	12	21
	1	2	3	4	5	9	12	21

This simple algorithm requires $O(n)$ exchanges and $O(n^2)$ comparisons. Its performance by no means matches that of insertion sort or quicksort. However, certain applications do not require all the elements to be sorted, and elements are needed on demand until a solution is found (see discussion on priority queues below). If the first k elements are demanded, this approach would need a time of $O(kn)$. This solution is satisfactory if k is usually small or n (the size of the list) is small. For general cases, heapsort, an optimised form of selection sort, can be used which requires a time of $O(n \log n)$ to sort the entire n elements.

Heapsort

In chapter 6 we introduced the heap data structure as an effective implementation of priority queues. The computations of priority queues and selection sort are analogous in the sense that both repeatedly find and remove the minimum element of a list.

Therefore in a list L stored as a heap, a sequence of n deletemin operations would retrieve all the elements in sorted order. To turn this simple scheme into a sorting algorithm, two problems should be tackled:

(1) As an element is deleted, where should it be stored?
(2) The original list should be turned into a heap.

The first problem can be solved by noting that in a heap $A[1] .. A[n]$, after one deletemin operation, the heap would be $A[1] ... A[n-1]$. The last element $A[n]$ is left unused. We can use it to store the minimum element of the list $A[1] .. A[n]$ (that is, $A[1]$). This, of course, means that the final list will be in non-increasing order. If a non-decreasing list is required, instead the heap can be organised to give the maximum element each time. Thus an algorithm can be developed for heapsort:

```
Heapsort:
    (* turn list L into a heap *)
    for i := n downto 2 do
        begin
            swap (1, i);
            (* Exchange A[1] with the last element A[i]. *)
            push (1, i-1)
            (* Push element A[1] into the sub-array
                A[1] .. A[i-1] to create a heap.
                See deletemin of chapter 6 and exercise 13 *)
        end;
```

Problem (2) above can be tackled by taking each element and inserting it in an initially empty heap. This, however, requires a separate array to hold the heap. A solution that avoids this extra doubling of storage starts from the middle of the list and pushes each element into the array ensuring that

$$a_i \leqslant a_{2i} \text{ and } a_i \leqslant a_{2i+1}$$

Such a 'heapify' operation would then be

```
Heapify : for i := n div 2 downto do
               push (i, n);
```

Let us look at our example list where $n = 8$. In heapify, the arrows indicate the elements which are to be pushed into the array and the elements marked with an asterisk are the children of those elements. In the sorting phase, when pushing an element, the same notation is used as in heapify. When swapping elements, the first and last elements are indicated by arrows.

	1	2	3	4	5	6	7	8
Array A:	4	12	3	9	1	21	5	2

$\uparrow$ *

```
Heapify:  push(4, 8)

          push(3, 8)      4     12     3      2      1     21     5    9
                                       ↑                    *     *
          push(2, 8)      4     12     3      2      1     21     5    9
                                ↑             *      *
          push(1, 8)      4     1      3      2      12    21     5    9
                          ↑     *      *
                          1     4      3      2      12    21     5    9
                                ↑             *      *
                          1     2      3      4      12    21     5    9
                                              ↑                        *
Heapified list:           1     2      3      4      12    21     5    9

Sorting:  swap(1, 8)      1     2      3      4      12    21     5    9
                          ↑                                            ↑
          push(1, 7)      9     2      3      4      12    21     5 |  1
                          ↑     *      *
                          2     9      3      4      12    21     5 |  1
                                ↑             *      *
          swap(1, 7)      2     4      3      9      12    21     5 |  1
                          ↑                                       ↑
          push(1, 6)      5     4      3      9      12    21 |   2    1
                          ↑     *      *
          swap(1, 6)      3     4      5      9      12    21 |   2    1
                          ↑                                ↑
          push(1, 5)      21    4      5      9      12 |  3      2    1
                          ↑     *      *
                          4     21     5      9      12 |  3      2    1
                                ↑             *      *
          swap(1, 5)      4     9      5      21     12 |  3      2    1
                          ↑                          ↑
          push(1, 4)      12    9      5      21 |   4     3      2    1
                                *      *
          swap(1, 4)      5     9      12     21 |   4     3      2    1
                          ↑                   ↑
          push(1, 3)      21    9      12 |   5      4     3      2    1
                          ↑     *      *
          swap(1, 3)      9     21     12 |   5      4     3      2    1
                          ↑            ↑
          push(1, 2)      12    21 |   9      5      4     3      2    1
                          ↑     *
          swap(1, 2)      12    21 |   9      5      4     3      2    1
                          ↑     ↑
          push(1, 1)      21 |  12     9      5      4     3      2    1

SORTED LIST:              21    12     9      5      4     3      2    1
```

The time requirement for heapsort consists of time for the heapify process and time for the sorting phase. Heapify performs $n/2$ push operations. In the worst case, each push requires $O(\log n)$ and therefore heapify needs $O(n \log n)$. (A closer analysis shows that heapify actually needs $O(n)$ – see exercise **9.14**). The sorting phase performs $n-1$ push operations and this in the worst case needs $O(n \log n)$. The total worst-case time complexity of heapsort is therefore $O(n \log n)$. However, the constant factor in this case is considerably larger than that of quicksort.

The problem of retrieving k elements in sorted order now requires $O(k \log n + n)$ and thus for $k \leqslant n/\log n$ the time requirement would be $O(n)$ (see exercise **9.15**).

9.4 Sorting by Utilising Representation of Elements

In the foregoing sections we assumed that our data type was a list of elements together with a relation '$\leqslant$'. That is, we could effectively *compare* any two elements and assert which one is smaller than the other one (or if they were equal). In this section we shall examine a sorting algorithm which does not use comparisons of elements. Instead it assumes that each element of the list is a string of symbols. Then, using the way these symbols are represented, we can sort the elements. For instance, for the following list

L: 123   018   218   114   110   311   121   016   236

each element is a string of three decimal digits. Each symbol can therefore have ten different representations.

This assumption implies that the elements are in a certain range. For example, in the above list, elements can only be in the range 000 to 999. Before we can generalise these ideas, let us take the above example and sort them using representations of decimal digits.

Pass 1 (a) Distribute the elements of L in that order in 10 queues using the least significant decimal digits:

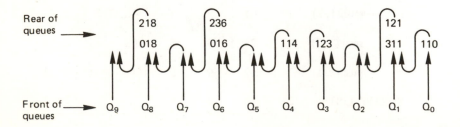

(b) Concatenate the 10 queues by joining the rear of queue Q_i to the front of queue Q_{i+1} (see the arrows in the above diagram). Thus the new list L constructed in this way will be

L: 11<u>0</u> 31<u>1</u> 12<u>1</u> 12<u>3</u> 11<u>4</u> 01<u>6</u> 23<u>6</u> 01<u>8</u> 21<u>8</u>

Note that the elements are sorted according to the least significant decimal digit.

Pass 2 (a) Distribute the elements of L in that order in 10 queues using the middle decimal digits:

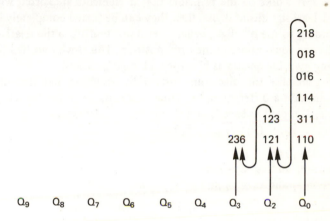

(b) Concatenate the 10 queues by joining the rear of queue Q_i to the front of queue Q_{i+1}. Thus the new list L constructed in this way is

L: 1<u>1</u>0 3<u>1</u>1 1<u>1</u>4 0<u>1</u>6 0<u>1</u>8 2<u>1</u>8 1<u>2</u>1 1<u>2</u>3 2<u>3</u>6

Note that at the end of this phase the elements are sorted with respect to the two least significant digits.

Pass 3 (a) Distribute the elements of L in that order in 10 queues using the most significant digit.

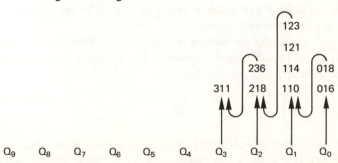

(b) Concatenate the queues to form L:

L: <u>016</u> <u>018</u> <u>110</u> <u>114</u> <u>121</u> <u>123</u> <u>218</u> <u>236</u> <u>311</u>

At the end of the third phase the elements are fully sorted. Note that no comparisons of elements were necessary.

In the above example the elements were in the range 0 to $10^3 - 1$; three passes were needed and ten queues were used. In general, if elements are in the range 0 to $k^p - 1$, then p passes and k queues will be needed. This corresponds to the situation where elements are p-digit numbers in base (or *radix*) k. This sorting algorithm is known as *distributive* sort, *bin* sort or *radix* sort. Primarily, distributive sort works on the principle that if elements are sorted with respect to the $p-1$ least significant digits, then they can be sorted completely by sorting them according to the pth digit, being careful not to disturb the relative order of elements having equal values in their p^{th} position. This is why we had to preserve this ordering by using queues as bins for each digit position.

We can generalise this algorithm further by assuming that each digit of the elements can have a different radix. Thus, elements are in the form $(e_1, e_2, \ldots, e_p)$ where e_i is a digit in base b_i and elements are in lexicographical order – that is, $(x_1, x_2, \ldots, x_p) < (y_1, y_2, \ldots, y_p)$:

> **if** $(x_1 < y_1)$ **or**
> $(x_1 = y_1$ **and** $x_2 < y_2)$ **or**
> $(x_1 = y_1$ **and** $x_2 = y_2$ **and** $x_3 < y_3)$ **or**
>
> .
>
> .
>
> .
>
> $(x_1 = y_1$ **and** $\ldots$ **and** $x_{p-1} = y_{p-1}$ **and** $x_p < y_p)$

In this form, p passes are required and in pass i, b_i queues are necessary. This generalised form is useful for sorting composite elements. For instance, consider the composite element type 'date' consisting of three components:

> day: 1..31
> month: 1..12
> year: 1900..1999

Sorting proceeds by three passes where pass 1 uses 31 queues, pass 2 uses 12 queues, and pass 3 uses 100 queues.

The time complexity for sorting n elements can therefore be calculated as $O(pn + S)$ where $O(pn)$ is used for distributing n elements p times and S is the time for concatenating queues in all passes.

$$S = O\left(\sum_{i=1}^{p} b_i \right)$$

In simple cases where $b_i = d$ for all i and for a given base d, the time requirement is $O(pn + pd)$. If p is small and independent of n, this is reduced to $O(kn)$ (see exercise **9.19**). This is the only sorting algorithm with a linear running time in terms of the size of the list.

The space complexity of the distribute sort is primarily that of queue headers and pointers for maintaining queue elements. The maximum number of queue headers is

$$O\left(\underset{i=1 \text{ to } p}{\text{MAX }} [b_i] \right)$$

Also, $O(n)$ pointers are needed to maintain queues. Therefore, this algorithm is really appropriate if the original list is represented as a linked list, in which case no extra pointers are necessary.

One advantage of the distributive sort is that the elements are not moved and only pointers are manipulated. This makes it suitable when the elements are very large, composed structures. Indeed, if each element is composed but small enough to fit into one word of memory, it is best to use one of the sorting algorithms discussed in previous sections.

9.5 Taxonomy, Comparisons and Related Issues

In section 9.1 we introduced an abstract sorting algorithm based on the divide and conquer approach. Furthermore, we divided the class of such algorithms into easy split/hard join and hard split/easy join. Each such sub-class was further divided according to whether splitting a list produced two equal-sized sub-lists or a singleton list and another list. We can now draw a map of this taxonomy – see figure 9.4. These algorithms are all based on the comparison and exchange of elements (if necessary). The distributive sort does not use this framework and is based on the representation of elements.

The graphs of figure 9.5 show comparisons of running times for different sorting algorithms. Three cases are considered: (a) ordered lists, (b) reversely ordered lists and (c) randomly ordered lists.

There are a couple of interesting points to note:

(1) For smaller lists, $O(n \log n)$ algorithms are worse than $O(n^2)$ algorithms.
(2) The rate of increase of time as the list size doubles is clearly visible for a linear algorithm (for example, insertion with an ordered list), a quadratic algorithm (for example, selection sort) and for logarithmic algorithms (mergesort, quicksort and heapsort).

These times were obtained on an Apple Macintosh computer. To avoid stack overflow, the quicksort algorithm was transformed into an iterative one by using an explicit stack. The lists were all lists of integers. The times for sorting large and composed objects would obviously be proportionally larger. In cases where the elements are very large, it may be appropriate to create a list of pointers to

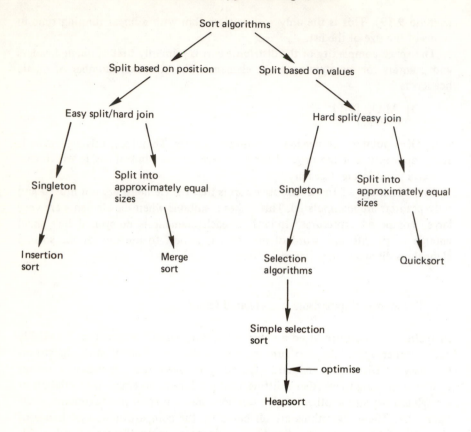

Figure 9.4 Taxonomy of sorting algorithms.

these items so that only pointers are changed instead of exchanging the actual elements.

Before we end this chapter on sorting algorithms, let us discuss a related problem to sorting, namely that of *order statistics*. In a list of *n* elements it is required to find the kth smallest (or largest) element of the sequence. For instance, if $k = 1$, $k = n$ or $k = n$ **div** 2, the minimum, maximum or the median of the list will be retrieved respectively.

The simplest solution would be to sort the list completely and then choose the kth element of the list. However, this requires $O(n \log n)$ in the best case. The heapsort could be used to find the first k elements (of the sorted list), and then to choose the kth element. We have shown that this would need a time of $O(n)$ if $k \leqslant n/\log n$.

For larger arrays, a variation of quicksort can be used which would run in an average time of $O(n)$. The list L (*n* elements) can be split into two sublists, L1 and L2, in the usual manner. Now, if $k \leqslant$ size(L1) then obviously the kth smallest

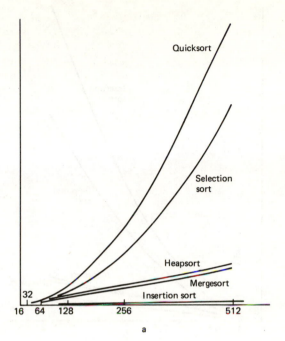

a

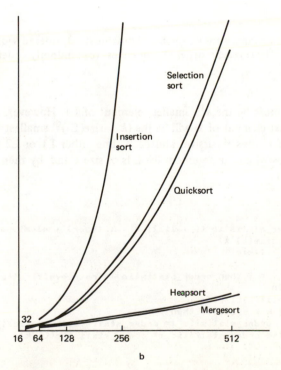

b

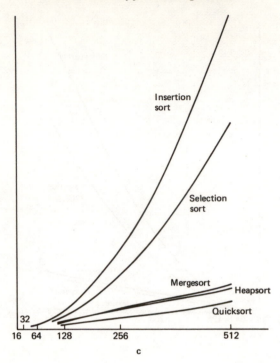

Figure 9.5   Comparisons of sorting algorithms: (a) ordered sequences;
                 (b) reversely ordered sequences; (c) randomly ordered sequences.

element of L must be the kth smallest element of L1. However, if $k >$ size(L1) the kth smallest element of L will be the $(k -$ size(L)$)^{th}$ smallest element of L2. The algorithm is thus recursive, and each time either L1 or L2 is chosen. The algorithm stops when the argument list L is of size 1 and, by then, obviously the argument $k = 1$.

```
function order_statistic (L : list; k : integer) : elemtype;
  (* 0 < k ≤ size(L) *)
  var L1, L2 : lists
  begin
    if size(L) = 1 then order_statistic := retrieve(first(L),L)
    else begin
         split(L,L1,L2);
         if k > size(L1) then
             order_statistic := order_statistic(L2,k-size(L1))
         else order_statistic := order_statistic(L1,k)
       end
  end;
```

The mechanics of this function are illustrated in the following example. Find the 5th smallest element of the list

	3	8	9	1	6	4	−1

$k = 5$, pivot $= 8$ 3   −1   4   1   6 | 9   8
$5 \leqslant 5$

             L1

$k = 5$, pivot $= 3$ 1   −1 | 4   3   6
$5 > 2$

             L2

$k = 3$, pivot $= 4$ 3 | 4   6
$3 > 1$

             L2

$k = 2$, pivot $= 6$ 4 | 6
$2 > 1$

             L2

$k = 1$, size(L2) $= 1$

Therefore, the 5th smallest element is 6. The time requirement of order_statistic can be shown to be $O(n)$ on average for any k (its analysis is similar to that of quicksort).

Exercises

9.1. Sort the following list of numbers using
   (a) merge sort
   (b) insertion sort
   (c) selection sort
   (d) heapsort
   (e) quicksort
   (f) distributive sort.

   25 2 36 142 101 84 32 49 8 321 25 458 981 122 48 95

9.2. Complete the merge sort algorithm by writing the procedure 'join'.
9.3. In insertion sort, assume that an extra element A[0] := −maxint exists. How does this simplify the algorithm?
9.4. Shell sort is defined as follows:

   • choose a d, an integer in the range $1 < d < n$
   • sort the d lists $a_1 \quad a_{1+d} \quad a_{1+2d} \cdots$
                  $a_2 \quad a_{2+d} \quad a_{2+2d} \cdots$

.

.

.

$$a_d \quad a_{2d} \quad a_{3d} \quad \ldots\ldots$$

using insertion sort
- choose a new d such that new $d <$ old d
- repeat the above until $d = 1$

Using shell sort, sort the list of exercise **9.1** using

initial $d = \log n + 1$
new $d =$ old d **div** 2

9.5. In insertion sort, in pass i, the proper place of a_i can be found by a binary search operation of the list $a_1 \ . \ . \ a_{i-1}$. How does such an improvement affect the overall complexity of the algorithm?

9.6. What does the following algorithm do?

```
procedure bubble;
   var i,j : integer;
   for i := 1 to n - 1 do
      for j := 1 to n - i do
         if A[j] > A[j+1] then swap(j,j+1);
```

9.7. What is the time complexity of the algorithm of exercise **9.6**?
       How can the algorithm in exercise **9.6** be improved?
       Hint: investigate the case when the list is already sorted.

9.8. *Tree sort*
       Using the elements of a list, a binary search tree can be constructed whose elements can subsequently be sorted by an in-order traversal. What is the time and space complexity of tree sort? What are the data dependencies of tree sort?

9.9. In quicksort, what happens if all the elements are equal? How can the algorithm be modified to optimise this special case?
       What if the pivot value is the median of the first, middle and last elements?

9.10. In quicksort, if we choose a single element (for example, the first element of the list as the pivot value) what will happen with our version of quicksort?

       Hint: consider an ordered list.

9.11. What is the space complexity of quicksort (best case, worst case and average case)?

9.12. In quicksort, we can apply order_statistic(L, n **div** 2) to get the median of the list L and use it as the pivot value. Is there any gain in doing so?

9.13. Write the push(i,u) procedure of heapsort.

9.14. (a) Show that

$$\sum_{i=1}^{m} \frac{i}{2^i} = 2 - \frac{m+2}{2^m}$$

(b) Show that 'heapify' requires a time of only $O(n)$.

Hint: Each a_i $\frac{n}{4} < i \leqslant \frac{n}{2}$ needs 1 comparison

Each a_i $\frac{n}{8} < i \leqslant \frac{n}{4}$ needs 2 comparisons

.

.

.

Each a_i $\frac{n}{2^{k+1}} < i \leqslant \frac{n}{2^k}$ needs k comparisons

9.15. Using the analysis of exercise **9.14**, show that to get the first k sorted elements of a list requires a time of $O(k \log n + n)$.
   For what values of k would this time be linear?

9.16. What are the best-case and average-case complexities of heapsort?

9.17. How does the bubble sort of exercises **9.6** and **9.7** fit in the taxonomy of section 9.5?

9.18. Investigate what sorting algorithms apply to the linked list implementation of lists.

9.19. On a computer with 16-bit integers we can sort a list of $n = 2^{16}$ elements using a distributive sort by using 16 passes and 2 queues (this assumes that positive integers are used). Is it true that this algorithm needs a time of $O(16 \times n + 16 \times 2) = O(n)$?

Bibliographic Notes and Further Reading

The subject of sorting is studied in detail by Knuth (1972). The conventional taxonomy of sorting algorithms, as in Knuth (1972), is based on low-level operations on the elements, namely insertion, exchange and selection.

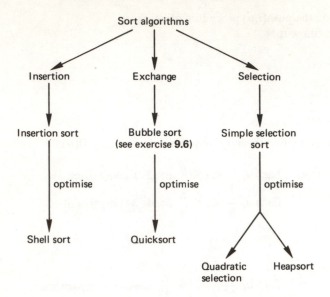

The bubble sort is a simple $O(n^2)$ algorithm. The quadratic selection is a $O(n\sqrt{n})$ algorithm (see Knuth (1972), p. 142). The taxonomy given in this chapter is a top-down approach and follows from an abstract algorithm on the ADT list. Other taxonomies are suggested by Darlington (1978) and Merritt (1985).

The shell sort of exercise **9.4** can be found in Knuth (1972) and is a $O(n^{1.5})$ algorithm. The quicksort is from Hoare (1962). The algorithms which use comparisons of elements are shown to require a minimum time of $O(n \log n)$. The discussion of this lower bound on sorting algorithms can be found in Aho *et al.* (1974).

Aho, A. V., Hopcroft, J. E. and Ullman, J. D. (1974). *The Design and Analysis of Computer Algorithms*, Addison-Wesley, Reading, Massachusetts.

Darlington, J. (1978). 'A synthesis of several sorting algorithms', *Acta Informatica*, Vol. 11, pp. 1-30.

Hoare, C. A. R. (1962). 'Quicksort', *Computer Journal*, Vol. 5, No. 1, pp. 10-15.

Knuth, D. E. (1972). *The Art of Computer Programming. Volume 3: Sorting and Searching*, Addison-Wesley, Reading, Massachusetts.

Merritt, S. M. (1985). 'An inverted taxonomy of sorting algorithms', *CACM*, Vol. 28, No. 1, pp. 96-99.

10 Graph Traversals and Algorithms

In this chapter we will discuss two systematic and structured methods of traversing the nodes and arcs of a graph. These traversal techniques can then be used as a powerful algorithm design tool on graph data types. These are indeed generalisations of the tree traversal methods which were used as a basis of efficient algorithms on trees.

10.1 Depth First Search

Depth First Search (dfs) of a graph $G = (V, E)$ is a generalisation of the pre-order traversal of a tree. Each node and arc is to be visited exactly once. Initially all the nodes are marked as 'unvisited'. Each unvisited node is then taken in turn, marked as visited and its unvisited adjacent nodes are in turn traversed in this same manner. An abstract formulation of dfs search is therefore

```
        for all v in V do declare v as unvisited;
        for all v in V do                        .......... (1)
          if v is unvisited then dfs(v);

where:  dfs(v) : mark v as visited;
                 for all (v,w) in E do           .......... (2)
                   if w is unvisited then dfs(w)
```

Since a node adjacent to v, say w, is searched before all other adjacent nodes to v, this sequential order of visiting (and, if needed, processing of) nodes and arcs is called depth first.

In the worst case, the **for** loop (1) looks at each node exactly once and the **for** loop (2) looks at each arc of the graph exactly once, and therefore the above search algorithm is $O(\max(n, e))$. However, since usually $e > n$, this would be simply $O(e)$.

Let us look at the graph G1 of figure 10.1a. A dfs search could start with node 1. Node 1 is visited and is given a label 1 indicating the order in which nodes are visited (see figure 10.1b). Nodes 2 and 4 are adjacent to node 1. Node 2 is visited with label 2. Search at node 2 is complete since it has no adjacent nodes. Node 4 should then be visited with label 3. The only adjacent node to 4 (that is, 2) is already visited and therefore the search at node 4 is complete. This in turn completes searching at node 1. Now another unvisited node should be

287

selected (say node 3). Node 3 is visited and labelled as 4. Its adjacent nodes 1 and 4 are already visited. Node 5 is then visited with label 5. Node 5 has nodes 6 and 7 next to it. Node 6 is taken first and is labelled 6. Nodes 3 and 4 are adjacent to 6, but both are already examined. The second adjacent node to 5, node 7, is labelled 7 and its only adjacent node 4 is visited. This completes the search at 5. Returning to the adjacent nodes of 3, node 7 is already visited and need not be searched again. The whole dfs search is now complete since no unvisited node remains. Nodes 1 to 7 are visited in the order 1, 2, 4, 3, 5, 6 and 7. This ordering of nodes is called *dfs ordering* of a graph. In figure 10.1b the arcs leading to an unvisited node are shown in solid lines and all other arcs are shown in broken lines.

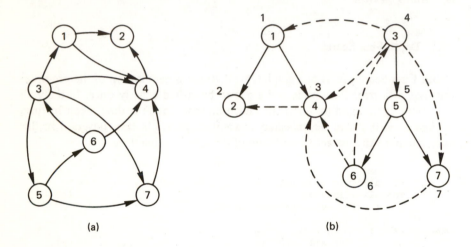

(a)   (b)

Figure 10.1   Digraph G1 and its dfs search: (a) graph G1 ; (b) dfs search of G1.

The solid arcs in a dfs search form a set of trees. Each such tree is a *depth first spanning tree* (dfst) and the solid arcs are called *tree arcs*. The set of these trees is referred to as a *depth first spanning forest* (dfsf). The dotted lines can also be classified into three different categories: *back arcs* are arcs from a node to an ancestor node in the same dfst; *forward arcs* are arcs from a node to a descendent node in the same dfst; all other dotted arcs are *cross arcs*. In the case of the dfsf of G1, arc (6, 3) is a back arc; arc (3, 7) is a forward arc; and arcs (3, 1), (4, 2) etc. are cross arcs.

The dfs can be equally applied to undirected graphs by taking each arc (*v, w*) as two arcs (*v, w*) and (*w, v*). The only differences are that forward arcs and back arcs are equivalent and no cross arc will exist in the dfsf of a graph (see figure 10.2).

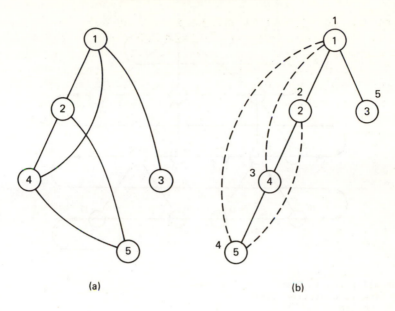

Figure 10.2 The dfs of an undirected graph G2: (a) graph G2; (b) dfst of G2.

10.2 Breadth First Search

In Breadth First Search (bfs) of a graph, the adjacent nodes (say w) to a given node v are visited and processed before visiting adjacent (and descendent) nodes of w. As in the depth first search, initially all the nodes are marked as 'unvisited'. Then, each unvisited node v is marked as 'visited', its adjacent unvisited nodes $w_1, w_2, \ldots, w_k$ (for some $k \geqslant 0$) are visited, then the adjacent nodes to $w_1, w_2, \ldots, w_k$, etc., are searched. In comparison with tree traversals, this is therefore a generalisation of the 'top-down left-to-right' search (see partially ordered trees of chapter 6). The diagram at the top of page 290 shows the order in which nodes of a graph are visited in bfs.

In the following bfs description, a queue of nodes (Q) is used to keep the visited nodes in the order of increasing levels of the nodes and from left to right within a level.

Figure 10.3 shows the bfs of graphs G1 and G2 of the previous section. For G1, node 1 is visited first. Its adjacent nodes 2 and 4 are visited and since they have no unvisited adjacent nodes the search at node 2 ends. The next unvisited node is node 3. After it has been visited, its unvisited nodes 5 and 7 are visited. Then the adjacent node to 5 (that is, node 6) is visited. Node 7 has no unvisited

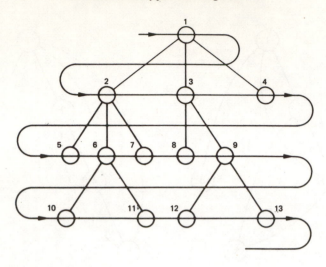

```
for all (v,w) in E do
   if w is unvisited then
      begin
         visit w;
         enqueue(Q,w)
      end;
   while not empty(Q) do        (* for all v in Q do bfs(v) *)
      begin
         bfs(front(Q));
         dequeue(Q)
      end
end;
```

The tail recursion in the above bfs algorithm can be removed by maintaining a single queue and a loop structure as

```
var Q : queue;     (* of node *)
   ....
   ....
initialise(Q);
for all unvisited node v do
   bfs(v);
```

where :-

```
bfs(v) :
   var w,x : node;
   begin
      visit v;
      enqueue(Q,v);
      while not empty(Q) do
         begin
            w := front(Q);
            dequeue(Q);
            for all (w,x) in E do
```

```
if x is unvisited then
    begin
        visit x;
        enqueue(Q,x)
    end
        end
end;
```

adjacent nodes, and the entire search is completed. The sequence of visiting the nodes is therefore 1, 2, 4, 3, 5, 7 and 6. Graph G2 is searched in a similar manner and figure 10.3b shows its bfs ordering of its nodes.

In a similar way to dfs, bfst and bfsf can be defined. Also, arcs in a bfst can be classified as tree, back and cross arcs. No forward arcs can exist. For un-directed graphs there will be only tree and cross arcs.

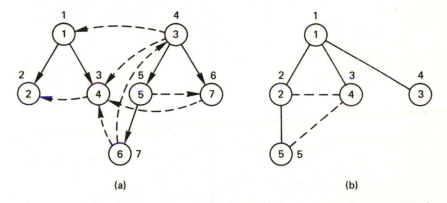

(a)   (b)

Figure 10.3   Breadth first search of graphs G1 and G2.

As an application of bfs on digraphs, we would mention Dijkstra's algorithm where a priority queue is used instead of a simple queue. A straight bfs would solve the shortest path problem (that is, when the weight of each arc is 1); see chapter 7, exercise **7.15**.

In the remainder of this chapter we shall take a look at some problems of graphs and show how these two search mechanisms can be used to design efficient algorithms for them.

10.3   Detection of Cycles

A common problem on graphs is to detect whether a given graph contains any simple cycles. For instance, in graph G1 (figure 10.1) (3, 5, 6, 3) is a simple cycle. As in figure 10.1b, in the depth first search graph, this cycle is represented by the tree arcs $(3, 5)$, $(5, 6)$ and the back arc $(6, 3)$. In general, if a graph G has a cycle $(v_0, v_1, v_2, \ldots, v_{k-1}, v_0)$ of length k, then it will be represented as $k-1$ tree arcs and one back arc in a depth first spanning tree. Let us assume v_i $(0 \leqslant i \leqslant k-1)$ is the first among these nodes to get visited. Because of the depth first behaviour of dfs, the search at node v_i will proceed as

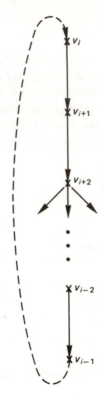

Each node may have other tree arcs incident on it (see v_{i+2}). The addition and subtraction of indexes above is to be interpreted in a cyclic manner (that is, **mod** k). To decide whether a graph G has a simple cycle, we can therefore perform a dfs and try to spot a back arc.

In the dfs algorithm of section 10.1, the tree arcs are very easy to identify. Simply, tree arcs are arcs (v, w) for which control reaches dfs(w) statement within the activation period of dfs(v).

The back arcs are from a node (v_{i-1}) to a proper ancestor (v_i) in the current dfst. Obviously, v_i must have been visited already. However, we cannot merely use the 'visited' markers to identify the back arcs since forward and cross arcs share the same property. First of all we must insist that v_i and v_{i-1} are in the same dfst. This would eliminate certain cross arcs from one spanning tree to another (for example, $(6, 4)$ in figure 10.1). Secondly, if v_i is the root of the current spanning tree being searched, we must ensure that v_i is a proper ancestor of v_{i-1} by restricting v_{i-1} to be on the *current search path* from the root v_i to this node v_{i-1}. This would then eliminate cross arcs such as $(4, 2)$ and forward arcs such as $(3, 7)$. In the former case the current search path would be $(1, 4)$ and 2 is not on the path. Similarly, in the latter example, the current path would be (3) and node 7 is not present on this path.

To implement this scheme we can use an extra marker for each node as a current search path – the *csp* indicator. When a node v is to be searched its csp is set to true and after all its descendants have been completely searched, its csp will be set to false. This enables us to detect back arcs (v_{i-1}, v_i) by checking if the csp of v_i is set to true. Note that if $csp(v)$ is found to be true, then v must have been visited already.

The search algorithm of section 10.1 can therefore be augmented as follows:

```
for all v in V do
   begin
      visited[v] := false;
      csp[v] := false
      (* v is not visited and is not on the current
         search path. It is assumed that 'node' is
         a subrange type and 'visited' and 'csp' are
         of type array[node] of boolean. *)
   end;

for all v in V do
   if not visited[v] then dfs(v);
where :
   dfs(v) : begin
               visited[v] := true;
               csp[v] := true;                    ..... (*)
               for all (v,w) in E do
                  if csp[w] then   A cycle is found   ..... (*)
                  else if not visited[w] then
                     dfs(w);
               csp[v] := false                     ..... (*)
            end;   (* of dfs *)
```

The extra statements added to the standard dfs algorithm are marked with asterisks (*). Visited and csp are mappings from nodes to a boolean value. For their implementations see the operation set_node_label of chapter 7 (section 7.2). Note that the time complexity of this algorithm remains at $O(e)$. This simple cycle-detection algorithm can be extended to output a complete cycle (if one is found) (a variant of this algorithm can be used to find all the cycles of a graph – see exercises **10.1** and **10.2**).

10.4 Topological Ordering

A Directed Acyclic Graph is known as a DAG. The topological ordering problem consists of ordering the nodes of a DAG G in a sequence such that if (v, w) is an arc of G then v appears before w in this sequence. Two simple realisations of this problem are prerequisite course structures and scheduling of programs in an operating system. In the former case, each course has a number of prerequisite courses which must have been taken before that course can be taken. The topological ordering amounts to the order in which all the courses can be taken such that the prerequisite requirements are not violated. In scheduling applications, each node is a process (program) and an arc (v, w) indicates that process v must be executed before process w.

These pieces of information can be represented by a DAG where, if course w pre-requires course v, then the arc (v, w) is in the graph. For instance, the graph of figure 10.4 is one such DAG.

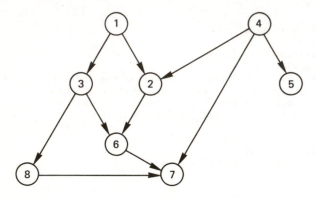

Figure 10.4   Example of a directed acyclic graph.

Course 7 requires courses 4, 6 and 8; course 4 has no prerequisites etc. One possible topological ordering of this DAG might be

    4   5   1   3   8   2   6   7

The topological ordering problem can be solved by using a depth first search of the DAG. Let the *successor* of a node v be all the nodes which can be reached from v via a path in the graph. In other words there is a path $(v, v_1, v_2, \ldots, v_k, w)$ from node v to a successor node w $(k \geq 0)$. According to the definition above, in the solution, node v_k must appear before w, node v_{k-1} must appear before $v_k, \ldots$, node v must appear before v_1. The linear structure of this sequence implies that v must appear before w for any successor node w.

In a depth first search of a DAG, when the search at a node v is completed, all its successors must have been visited by then. A dfs of our example above would be

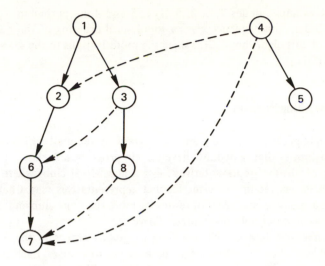

For each node its adjacent arcs are either tree, cross or forward arcs (no back arcs can be present in a DAG). These arcs, however, have already been visited and depth first strategy ensures that all the successor nodes have already been visited. To derive a topological ordering, therefore, when the search at a node v is completed, we add it before all its successors in the sequence. We do this by pushing such visited nodes onto a stack. A sequence of pop operations will then give us the required ordering:

```
    initialise(stk);          (* stk is a stack of nodes *)
    for all v in V do visited[v] := false;
    for all v in V do
        if not visited[v] then
            dfs(v);
    while not empty(stk) do
        begin
          write(top(stk));
          pop(stk)
        end;
where :-
    dfs(v) : begin
                visited[v] := true;
                for all (v,w) in E do
                    if not visited[w] then
                        dfs(w);
                (* At this stage all the descendants of v are pushed
                   into the stack. So we can now push v into the stack *)
                push(v,stk);
             end;
```

In the above example, nodes 7, 6, 2, 8, 3, 1, 5 and 4 are pushed into the stack stk in that order. This sequence gives a topological ordering of the nodes. The complexity of this algorithm is $O(e)$ since the modifications to the dfs algorithm are minimal ($O(1)$).

10.5 The Matching Problem

In many graph problems, a sub-graph of a graph is required which *matches* a particular pattern — that is, it must have certain properties. A classic example is the problem of 'distinct representatives'. For a number of clubs where each has many members, we require to elect distinct representatives (say Chairpersons etc); that is, a representative cannot represent more than one club and each club can have a maximum of only one representative.

Another manifestation of the matching of graphs occurs in the 'lecturers-courses' problem where lecturers are to be assigned to courses. Each course can be taught by a number of lecturers. Indeed, this problem is an instance of the 'distinct representative' problem.

In abstract form, we can represent this matching problem as a bipartite graph.

Definition. A graph G = (V, E) is bipartite if there exists two *disjoint* subsets of V, say V_1 and V_2 such that $\nexists$ $(x, y) \in E$ such that (x and $y \in V_1$) **or** (x and $y \in V_2$).

For example

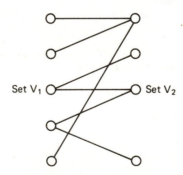

Definition. A subset of E, E′ with no two edges incident upon the same vertex in V is called a *matching*. A matching with the maximum number of edges in it is called a *maximal matching*.

Definition. A *complete* matching is one such that every vertex in V is an end of some edge in the matching.

Definition. In a matching M, every vertex and arc in M is said to be *matched*. Other nodes and arcs of the graph G are said to be *unmatched*.

If a complete matching exists for a graph G, then $|V_1| = |V_2| = \alpha$ and $|E'| = \alpha$.

Now, returning to the distinct representative problem, a bipartite graph can represent the problem where clubs $\equiv V_1$, members $\equiv V_2$.

Since, in general, $|V_2| \gg |V_1|$, we redefine a complete matching as follows.

Definition. A complete matching E' is one which has one edge incident upon each vertex in V_1. Therefore $|E'| = |V_1|$.

So in abstract form the problem of distinct representatives reduces to finding a complete matching in a bipartite graph. Of course, there may not be a solution if $|V_2| < |V_1|$.

There are many solutions to this problem. In a simple solution we can find all matchings and then pick the maximum one. However, this will lead to an exponential algorithm. A better solution is based on *Augmenting Paths* (AP).

Definition. An augmenting path G relative to a matching M is a path connecting two unmatched vertices in which alternate edges are in M.

Now, given a matching M and an augmenting path G relative to M, we can construct a larger matching. First we observe that an AP is of odd length and its first and last edges are not in M. A larger matching can be obtained by forming the set $(M \cup G) - (M \cap G)$ — that is, the edges which are in M and G but not both. Let us denote this set operation as $M \oplus G$.

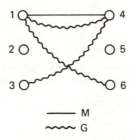

Now $M \oplus G$ is

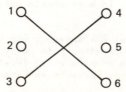

In fact, we can show that $|M \oplus G| = |M| + 1$. Therefore $M \oplus G$ is a larger matching than M.

For our problem, a matching M is complete if $|M| = |V_1|$. This implies that when $|M| = |V_1|$, there cannot be an AP relative to M. So the algorithm for finding a maximal matching is as follows:

1. M := { }
2. Find an AP G relative to M
3. Replace M by M ⊕ G
4. Repeat 2 and 3 until no more AP can be formed.

Step (2) in this algorithm is the crucial part and we now show a depth first search that can be used to find an AP.

Since G is a bipartite sub-graph, we can proceed incrementally, starting from a vertex in V_1 and alternately go to V_2 and V_1. That is, we build an AP for levels i, where $i = 1, 2, \ldots$ using a process rather like depth first search. At $i = 1$, we begin with a new unmatched vertex in V_1. At even level i, we visit new vertices that are adjacent to a vertex matched at level $i-1$, by a non-matched edge (that is, one which is not in M) and we also include that edge in G. At odd level i, we visit new vertices that are adjacent to a vertex at level $i-1$ because of a matched edge in M, and include that edge in G. This process is repeated until an unmatched vertex in V_2 is added at an even level *or* no more vertices can be visited. If no vertices can be added to the current incomplete AP being formed, the dfs should look for another AP (by backtracking).

Let us assume that each edge and node has a marker 'matched' to indicate whether or not it is in a matching M. We can now describe the algorithm for finding an augmented path as follows:

```
M := {};
for all a in E do declare arc a as not matched;
for all v in V do declare node v as not matched;
for all v in V₁ do
    if v is not matched then
        begin
            G := {};
            dfs(v,1,found);
            if found then
                M := M ⊕ G (* appropriate matched markers are set *)
        end;
where :-
    procedure dfs(v : node; L : integer; var found : boolean);
        (* using a dfs an augmenting path relative to M is to
            be found starting at node v, at level L. If one such
            AP is found, a true value is returned in 'found' *)
        begin
            found := false;
            if (L is even) and (v is not matched) then found := true
            else for all a≡(v,w) in E do
                    if not found then
```

```
     if ((a is matched) and (L is even)) or
        ((a is not matched) and (L is odd)) then
           begin ,
              dfs(w,L+1, found)
           if found then
                 insert(a,G)
           end
end; (* of dfs *)
```

The operation of this algorithm is shown below for the following bipartite graph:

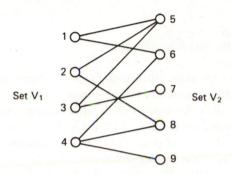

where $V_1 = \{1,2,3,4\}$ and $V_2 = \{5,6,7,8,9\}$.

M	dfs(v,1,found)	G	M ⊕ G	Matched nodes
{ }	1	{(1,5)}	{(1,5)}	{1,5}
{(1,5)}	2	{(2,5), (5,1), (1,6)}	{(1,6), (2,5)}	{1,2,5,6}
{(1,6), (2,5)}	3	{(3,5), (2,5), (2,8)}	{(1,6), (3,5), (2,8)}	{1,2,3,5,6,8}
{(1,6), (3,5), (2,8)}	4	{ (4,9) }	{(1,6), (3,5), (2,8), (4,9)}	{1,2,3,4,5,6,8,9}
{(1,6), (3,5), (2,8), (4,9)}				

Initially M is empty. A dfs at node 1 finds the unmatched node 5. Search at 5 terminates since level = 2 (even). Thus (1,5) is added to G and M is replaced by M ⊕ G. The next unmatched node in $\bar{V}_1$, node 2, is searched next. Arc (2,5) is chosen first. At level 2, arc (5,1) which is already matched is chosen. At level 3, arc (1,6) qualifies since node 6 is unmatched. The search then terminates and arcs (1,6), (5,1) and (2,5) will be added to G in that order. M ⊕ G now becomes

{(1,6), (2,5)}. A dfs at node 3 and another at node 4 eventually lead to the matching {(1,6), (3,5), (2,8), (4,9)}.

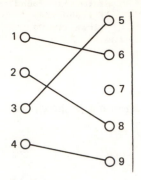

Implementation of the Matching Algorithm

The matching M and the augmenting path G can be represented as ordered lists of arcs. Therefore, the $\oplus$ operation can be executed in O(length M + length G). This is bounded by O(e). An alternative implementation would be to use a bit-vector of e bits to represent M and G. In this case the $\oplus$ operation would require O(e)-bit operations alone. The graph G can be implemented using multi-linked lists where each arc occurs in two adjacency lists of its two end nodes.

If $|V_1| + |V_2|$ is bounded by a constant N, then the following data structures can be used to implement a graph G and the matchings (see figures 10.5 and 10.6).

```
const N = ;     (* maximum number of nodes in G *)
type node = 1..N;
     eptr = ↑edge;
     node_cell = record
                    matched : boolean;
                    first : eptr
                 end;
     edge = record
               v,w : node              (* this arc corresponds to (v,w) *)
               matched : boolean;
               next_v, next_w : eptr (* to next arcs (v,-) and (w,-) *)
            end;
     cell_ptr = ↑cell;
     cell = record
               edge : eptr
               next : cell_ptr
            end;
     list = cell_ptr;
var nodes : array [node] of node_cell;
    G,M : list;
    (* ordered lists of arcs based on (v,w) components of arcs *)
    V_1,V_2 : 1..N;
```

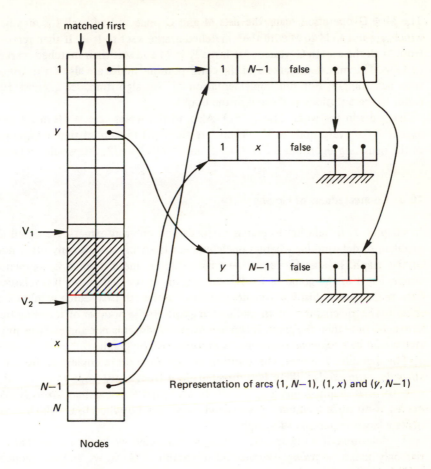

Figure 10.5   Data structures for implementing graph G.

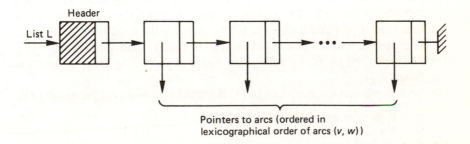

Figure 10.6   Implementation of matchings and augmenting paths.

The M $\oplus$ G operation scans the lists M and G once and adds all the arcs in G which are not in M to M with their matched markers set to true. It then removes from M all the arcs in G which were initially in M and sets their matched markers to false. The matched markers of the corresponding nodes are also set to true or false accordingly. For full implementation of this algorithm, see exercise **10.6** (refer to the Solutions at the end of the book).

Each dfs in the worst case is $O(e)$. Also, in the worst case, a dfs needs to be executed for every node in V_1. Therefore the total maximum time complexity of the matching algorithm is $O(|V_1| \times e)$ and if $|V_1| \approx n/2$, this would be $O(ne)$.

10.6 Connectedness of Graphs

In chapter 7, in relation to graphs of roads and one-way streets, we posed the problem of determining whether one could get from any node to any other node of the graph. In abstract terms, we wish to determine if a graph is completely connected. For the graph of roads, which is an undirected graph, it is relatively easy to determine this. Indeed a depth first or a breadth first search would determine the connectedness of an undirected graph. If the number of spanning trees is exactly one, then the graph is connected; otherwise it is not and each spanning tree would be a *maximal connected component* of the graph.

For digraphs, however, the solution is slightly more complex. A digraph is *strongly connected* if there is a directed path from every node to every other node of that graph. A sub-graph of G' of a digraph C is *maximal strongly connected* if no node v and arc a not already in G' can be added to G' which would make a larger connected sub-graph.

To determine if a digraph G is strongly connected we have to show that G is the only maximal strongly connected sub-graph of G. So we will now concentrate on this problem.

First, consider the problem of finding the maximal strongly connected sub-graphs of a digraph G. An algorithm for this problem can be developed as follows:

(1) Perform a dfs on G = (V, E) and label each node according to the topological ordering of the nodes. Cycles must be ignored (that is, ignore the back arcs).
(2) Construct a new graph G' = (V, E') by reversing the direction of all the arcs in E; that is, for each arc (v, w) in E, the arc (w, v) is added to E'.
(3) Perform a dfs on G' by taking the nodes in increasing order of labels as found in (1).
(4) Each depth first spanning tree of (3) is a maximal connected sub-graph of G.

Let us look at an example of a digraph G:

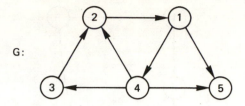

Note that the nodes {1,2,4} and {1,2,3,4} are connected sub-graphs, whereas only {1,2,3,4} is a maximal connected sub-graph.

The dfs of this graph and its topological ordering is

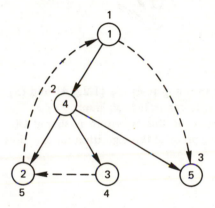

The digraph G' is therefore

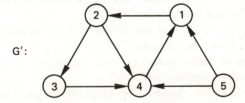

Now, we apply a dfs on G', starting with node 1 which has the lowest topological ordering label:

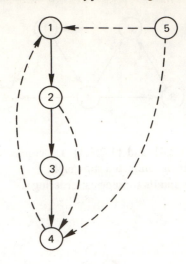

The two depth first spanning trees are {1,2,3,4} and {5}. Each such spanning tree is a maximal strongly connected sub-graph.

The time complexity of this algorithm is simply $O(2e)$ since the graph is searched twice. For the proof of this algorithm see exercises **10.11** and **10.12**.

Exercises

10.1. Write a cycle-detection algorithm which would print a cycle (as a sequence of nodes) when one was found.

10.2. Write a program to find all the simple cycles of a digraph. What is the time complexity of your program?

10.3. How can a bfs be used to detect cycles in undirected and directed graphs? What are the time complexities?

10.4. Show that a complete matching of a bipartite graph G must have $|V_1| = |V_2|$.

10.5. Show that $|M \oplus G| = |M| + 1$.

10.6. Write a program to read a bipartite graph and find a matching for it.

10.7. If d_v is used to indicate the order in which v was visited in a dfs of a graph, show that if an arc (v, w) is a cross arc, then $d_v > d_w$.

10.8. Expressions in Pascal can be represented as DAGs. For instance, the expression $(a \times b) + (2 - 3/(a \times b))$ can be represented as

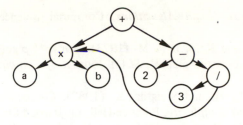

Write a program which would read an expression and construct its DAG.

10.9. *Hamilton cycle problem*

Write a program which detects if a graph has a cycle which passes through each node exactly once (for all nodes of the graph).

What is time complexity of your solution? (See also chapter 12.)

10.10. How can the cycle detection algorithm be used to solve the following problem:

"Is there a cycle of length k in a given digraph?"

What is the special case when $k = n$?

10.11. Show that in the maximal strongly connected algorithm of section 10.6, if there is a path from v to w and a path from w to v then v and w are in the same dfst of G'.

10.12. Show that in the algorithm of section 10.6, if two nodes v and w, are in the same dfst of G', then there must be a path from v to w and a path from w to v in G.

Hint: If u is the root of the dfst in G' show that label(u) $\geqslant$ label(v) and label(v) $\geqslant$ label(w). Then use the topological ordering properties of the labels.

Bibliographic Notes and Further Reading

Algorithmic graph theory is a well-studied branch of applied mathematics and for a wider exposition of the subject refer to Christofides (1975), Carre (1979), Evans (1980) and Papadimitriou and Steigtilz (1982).

The maximal matching algorithm of section 10.5 is due to C. Berge and J. Edmonds, and an $O(n^{2.5})$ for finding maximum matchings in bipartite graphs is given by Hopcroft and Karp (1973).

The connected sub-graph algorithm is from Shavir (1981).

Carre, B. (1979). *Graphs and Networks*, Oxford University Press.

Christofides, N. (1975). *Graph Theory: An Algorithmic Approach*, Academic Press, London and New York.

Evans, S. (1980). *Graph Algorithms*, Computer Science Press, Rockville, California.

Hopcroft, J. E. and Karp, I. R. M. (1973). 'An $n^{5/2}$ algorithm for maximum matchings in bipartite graphs', *SIAM Journal of Computing*, Vol. 4, No. 2, pp. 225–231.

Papadimitriou, C. H. and Steigtilz, K. (1982). *Combinatorial Optimization: Algorithms and Complexity*, Prentice-Hall, Englewood Cliffs, New Jersey.

Shavir, M. (1981). 'A strong connectivity algorithm and its application in data flow analysis', *Computer and Mathematics with Applications*, Vol. 7, No. 1, pp. 67–72.

11 String-searching Algorithms

In this chapter we introduce the abstract data type string and describe some basic string-processing operations. The purpose here is not to give a full exposition of special-purpose string-handling facilities, but instead to concentrate on string searching and emphasise the role of algorithm design. The hope is that these algorithms will further elicit the algorithm design paradigms discussed in chapter 8. More specialised string-processing systems and tools such as text editors and specialised string-manipulation languages such as SNOBOL offer powerful facilities. The interested reader is referred to the bibliographic notes at the end of the chapter.

11.1 Definitions

The abstract data type *string* is a sequence of characters. A string S consists of the sequence "$c_1 c_2 \ldots c_n$" ($n \geq 0$; $n=0$ signifies an empty string) where c_i is a character. As a basic data type the set of operations on strings can be summarised as

    procedure   create (**var** S : string);
                    This creates an empty string.

    procedure   single (c : char; **var** S : string);
                    This creates a string S containing the character c.

    procedure   append (S1, S2 : string; **var** S3 : string);
                    If S1 = "$c_1 c_2 c_3 \ldots c_n$" and
                      S2 = "$c'_1 c'_2 c'_3 \ldots c'_m$"
                    then S3 will be "$c_1 c_2 \ldots c_n c'_1 c'_2 \ldots c'_m$".

    procedure   copy (S1 : string; **var** S2 : string);
                    This is equivalent to S2 := S1.

    function   equal (S1, S2 : string) : boolean;

    function   index (S : string; n : integer) : char;
                    Function index returns the nth character of S.

    function   length (S : string) : integer;
                    This function returns the number of characters in S.

procedure   substring (S : string; i,j : integer; **var** SS ; string);
If $S \equiv$ "$c_1 c_2 c_3 \ldots c_n$" then this procedure assigns the
substring ($c_i c_{i+1} \ldots c_j$) to SS.
$1 \leqslant i \leqslant j \leqslant n$.
Note that: SS := index (S,k) is equivalent to
substring(S,k,k,SS).

function   search (t, p : string) : integer;
text   t $\equiv$ "$t_1 t_2 \ldots t_n$" t_i is a character
pattern p $\equiv$ "$p_1 p_2 \ldots p_m$" p_i is a character
Function search returns the *smallest k* such that

$$t_k \quad = p_1$$
$$t_{k+1} \quad = p_2$$
$$\ldots \quad \ldots$$
$$t_{k+m-1} = p_m$$

If there is no such k, a zero will be returned.

If desired, procedures insert(S1,S2,i) and delete(S,i,j) can be defined to insert
string S1 at position i of S2 and to delete string "$s_i \ldots s_j$" from S to abstract a
particular implementation being used.

Here are some examples:

append: S1="man' S2="made" S3="manmade"
index: S="computer' n=4 index='p'
length: S="strings" length=7
substring: S="this-is-a-string" i=3 j=12
           SS="is-is-a-st"
search: text="this-is-a-string-of-length-29"
           pattern="a-str"
           search=9

In certain applications two additional operations, first(S) and next(S), can be
defined. The former representation sets a 'currency' indicator to the beginning of
the string while the latter retrieves the next character after the current position
and updates the current position.

11.2   Data Structures for Strings

The simplest way to represent strings using data structures is to assume that all
the strings are of fixed length L (say) and use an array of characters:

```
const L = ;
type string = packed array [1..L] of char;
```

This, of course, implies that all the strings longer than L characters cannot be represented and furthermore, for strings shorter than L characters, their array representation should be filled with spaces. For instance, if L = 10, the string "man" would be stored as "man-------". Long strings will be truncated. For instance "examinations" would be stored as "examinatio".

A slightly improved data structure may insist on a maximum length for strings but will allow the actual length of any string to be dynamically determined up to the maximum length:

```
const max = ;
type string = record
              ch : packed array [1..max] of char;
              length : 0..max
          end;
```

Some versions of Pascal, such as UCSD and SVS, provide a pre-defined type *string* which is essentially what we have described above. In these systems, although the dynamic length of a string can vary from 0 to max, the storage is allocated for holding *max* characters.

In applications where there cannot be any maximum limit on the length of the strings, storage for strings must be allocated dynamically. This, of course, necessitates more complex allocation and de-allocation procedures to ensure the optimum use of storage. In the remainder of this chapter we shall generally assume that the above approach is used where a maximum limit is enforced on the size of the strings.

Implementing the string operations using the array structures is straightforward for all the operations, except perhaps for the search operation for which there can be more than one algorithm. We shall look at a few search algorithms in the next section.

11.3 String-searching Algorithms

In section 11.1 the function search was defined to find the *first* occurrence of a pattern string in a text string. Let p_i and t_i denote the ith characters of a pattern string p and a text string t where length(p) = m and length(t) = n. We shall use three algorithms for implementing this function. The first one is a Simple algorithm (S). The other two will utilise ideas from dynamic programming and lead to two table-driven algorithms.

Simple Algorithm (S)

The pattern p is aligned with the text string at its extreme left so that t_1 and p_1 are aligned:

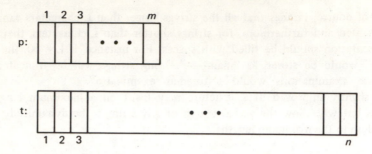

The pattern and text characters are then compared from left to right until either a match is found or a mismatch occurs. If a mismatch occurs, the pattern is shifted one position to the right and the scan is restarted as

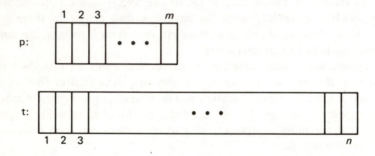

Let us use the variables i and j to index into the pattern and text respectively. Furthermore, let k denote the position in t which aligns with p_1. In general, when the first $i-1$ characters of the pattern are matched (the striped characters, see top of next page) and the ith character is mismatched against the text, i, j and k should be modified to 1, $k+1$ and $k+1$ respectively, and the search should restart as

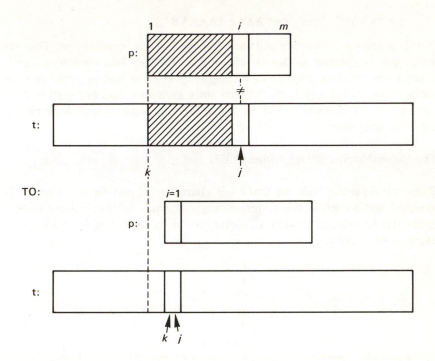

Algorithm S:

```
function search(t,p : string) : integer;
  label 99;
  var i,j,k,m,n : integer;
  begin
    m := length(p);  n := length(t);
    k := 0;  search := 0;
    while k < (n-m+1) do
      begin
        k := k + 1;
        (* the position in t aligned with position 1 in p *)
        j := k;    (* backtrack *)
        i := 1;
        (* search t, t_{j+1} ... and p_1 p_2 ... *)
        while (index[p,i] = index[t,j]) do  (* p_i = t_j *)
          if i = m then  (* a match is found *)
            begin search := k; goto 99 end
          else begin
                  j := j + 1  (* keep matching *)
                  i := i + 1
               end
      end
99 : end;  (* of search *)
```

The worst-case performance of algorithm S occurs when, for every possible starting position (*k*), all characters of the text except the last character match the pattern. For example p = "AAAB"   and   t = "AAAAAAAAAB"

In this, n scans are necessary and each scan performs m comparisons. Thus the worst-case complexity is $O(mn)$. The reason for this behaviour is primarily because the searching backtracks (through $j := k$). It should be noted that the average time complexity is much better since each scan does not need to compare all the m characters. The next two algorithms avoid such backtracking behaviour altogether.

The Knuth–Morris–Pratt Algorithm (KMP)

When a comparison fails, the first $i - 1$ characters of patterns are successfully matched and therefore the corresponding characters in the text are known. Using this knowledge, the index j never has to be backed up ($j := k$). Let us examine an example:

$$i = 5$$
$$\downarrow$$

```
p:  A   B   C   A   C
t:  A   B   C   A   B   C   A   C   A   B
```
$$\uparrow$$
$$j = 5$$

Algorithm S would restart the search at

$i = 1$	$j = 2$	$\ldots$	(I)
$i = 1$	$j = 3$	$\ldots$	(II)
$i = 1$	$j = 4$	$\ldots$	(III)

Since we know that the characters "$t_1\ t_2\ t_3\ t_4$" are the same as "$p_1\ p_2\ p_3\ p_4$" and that $t_5 <> p_5$, we can conclude that (I) and (II) must fail and (III) succeeds. Therefore, we can jump directly to $i = 2$ and $j = 5$:

$$i = 2$$
$$\downarrow$$

```
            A   B   C   A   C
A   B   C   A   B   C   A   C   A   B
```
$$\uparrow$$
$$j = 5$$

Thus, j does not change, but i is modified to a new position. If (III) was known to fail, the search would be restarted at $i = 1$ and $j = 6$ (the next position in text) as is the case with p = "ABCDC" and t = "ABCDBACDC".

In general, when a failure occurs, j is kept unchanged and i is changed to a new position fail $[i]$. By convention, if fail $[i] = 0$, then the search will be restarted at $i = 1$ and $j = j + 1$. Thus an outline of this strategy is

KMP algorithm:

```
    function search(t,p : string) : integer;
      ...
    begin
      ...
    j := 1;  i := 1;  search := 0;
    while (i <= n) and (j <= m) do
      if index[p,i] = index[t,j] then (* p  = t  *)
                                        i    j
        if i = m then                 (* a match is found *)
          begin search := j - m; goto 99 end
        else begin
              i := i + 1;             (* keep matching *)
              j := j + 1
            end
      else                            (* move pattern forward *)
        if fail[i] = 0 then
          begin
            i := 1;
            j := j + 1
          end
        else i := fail[i]
99 : end;  (* of search *)
```

To complete the KMP algorithm we need to compute the mapping fail[i] for all $1 \leqslant i \leqslant m$. This can be done by examining the pattern independently of the text string. The general principle is outlined in figure 11.1.

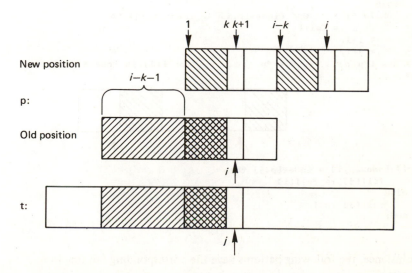

Figure 11.1 Computation of the fail function.

After a failure at position i of the pattern has occurred, the pattern can be shifted $i-k-1$ positions forward if the first k characters of the pattern and the last k characters before position i of the pattern are equal — that is, "$p_1 p_2 \ldots p_k$" = "$p_{i-k} p_{i-k+1} \cdots p_{i-1}$". To formalise this, we could specify the fail function as

```
Compute fail[i] :
    if for some k : k < i-1 and "P₁ ... Pₖ" = "Pᵢ₋ₖ ... Pᵢ₋₁"
            and k is the largest such value then
        if index[p,i] <> index[p,k+1] then
            fail[i] := k + 1
        else fail[i] := fail[k+1]
    else if index[p,i] = index[p,1] then fail[i] := 0
        else fail[i] := 1
```

If $p_i = p_{k+1}$, since $t_j <> p_i$, obviously t_j fails to match p_{k+1}. To prevent matching p_{k+1} and t_j, fail$[i]$ is set to fail$[k+1]$. The latter formulation is recursive. However, since fail$[x]$ $1 \leqslant x \leqslant i-1$ are known when computing fail$[i]$, the value of fail$[k+1]$ can be read directly from a table (an array $[1..n]$).

The following fragment of a Pascal program implements the above algorithm for computing the fail function:

```
i := 0   fail[1] := 0;
j := 1;
while j <= m do
    begin
        while (i > 0) and (index[p,i] <> index[p,j]) to
                i := fail[i];
        i := i+1;
        j := j+1;
    (* Now ("P₁ P₂ ... Pᵢ₋₁" = "Pⱼ₋ᵢ₊₁ ... Pⱼ₋₁" or i=1) is true *)
    (*
```

```
    *)
        if index[p,i] = index[p,j] then
            fail[j] := fail[i]
        else
            fail[j] := i
    end;
```

For instance, the following patterns have the corresponding fail function:

pattern:	A	B	C	A	C
fail:	0	1	1	0	2

pattern:	A	A	A	A	A	A	A	A	B
fail:	0	0	0	0	0	0	0	0	8

The above program to compute the fail function is of $O(m)$. The KMP search function is $O(n)$ since no character in the text is compared more than twice (in the worst case). Thus the total worst-case complexity is $O(m+n)$.

The Boyer–Moore Algorithm (BM)

Unlike the Simple and KMP algorithms where searching proceeds from the left of the pattern, the BM algorithm starts searching from the right of the pattern. In doing so, when a failure occurs, it is possible to shift the pattern further and consequently some text characters need not be considered in the search.

Initially, the text and pattern strings are aligned at their leftmost positions. The search begins by comparing character pairs $(p_m, t_m) (p_{m-1}, t_{m-1})$ and so on. If a mismatch occurs when comparing p_i and t_i (in general), a right shift of the pattern can be computed using two pre-computed functions fail1 and fail2.

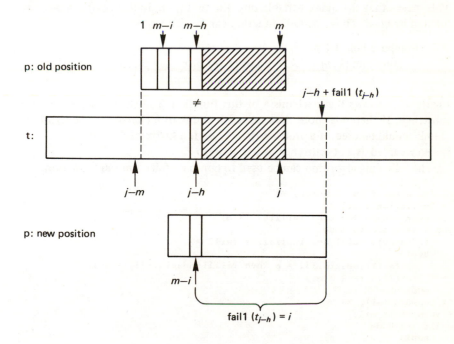

Figure 11.2 Computation of fail1 function.

The fail1 function is based on the idea that if a failure occurs at position $j-h$ in the text (see figure 11.2), the pattern can be shifted to the right such that

t_{j-h} matches a character p_{m-i}. Naturally we choose the smallest such i. Thus fail 1 is a function of a character (the failed character t_{j-h}):

Compute fail1 $[c]$:
  if c does not occur in the pattern **then** fail1 $[c]$ = m
  else fail1 $[c]$ = the smallest i such that $0 \leqslant i < m$ and $p_{m-i} = c$.
  or equivalently :
  fail1 $[c]$ = minimum $\{i \mid i=m$ **or** $(0 \leqslant i < m$ and $p_{m-i} = c)\}$

As in figure 11.2, in general the pattern is shifted such that the $m-i$ text characters following t_{j-m} are not examined.

The fail2 function is more like the fail function of KMP and is a function of the position in the pattern at which a failure occurs. If a mismatch occurs at position $m-h$ of the pattern, the pattern can be shifted to the right such that p_d aligns with t_{j-h} if "$p_{d+1} \cdots p_{d+h}$" = "$p_{m-h+1} \cdots p_m$" and $p_{m-h} \neq p_d$. In other words, there must be a chunk of h characters (S, say) in the pattern which is the same as the last h characters of the pattern. If there exists no such chunk of similar h characters, then the pattern can be shifted to the right by m positions. This means that the index variable into the text ($j-h$, in this case) can be incremented by $m-d$. Thus, the fail2 function can be defined as

Compute fail2 $[i]$ =
  minimum $\{k+m-i \mid k \geqslant 1$ **and** $[k \geqslant j$ **or** $p_{j-k} = p_j$; for $i \leqslant j \leqslant m]$
  and $[k \geqslant i$ **or** $p_{i-k} \neq p_i]\}$

Figure 11.3 shows the mechanism of this function. The striped areas are equal substrings. If there are more than one such chunk of h characters, the rightmost chunk should be used. If a prefix '$p_1 p_2 \ldots p_i$' is a suffix of S, then '$p_1 p_2 \ldots p_i$' is also regarded as an occurrence of S in pattern P.

The following algorithm can be used to compute fail1 and fail2 functions:

```
(* The array f is temporary *)
(* Initialise fail1 *)
for every character c do fail1[c] := m;
for j := m downto 1 do
    (* Compute fail1 and initialise fail2 *)
    begin
        if fail1[index[p,j]] = m then fail1[index[p,j]] := m-j;
        fail2[j] := 2 * m-j
    end;
(* Compute fail2 *)
j := m; t := m+1;
while j > 0 do
    begin
        f[j] := t;
        while (t <= m) and (index[p,j] <> index[p,t]) do
            begin
                fail2[t] := minimum (fail2[t], m-j);
                t := f[t]
            end;
        j := j-1; t := t-1
```

```
      end;
for k := 1 to t do
      fail2[k] := minimum (fail2[k], m+t-k);
tp := f[t];
while (t <= m) do
      begin
        while (t <= tp) do
            begin
                fail2[t] := minimum (fail2[t], tp-t+m);
                t := t+1
            end;
        tp := f[tp]
      end;
```

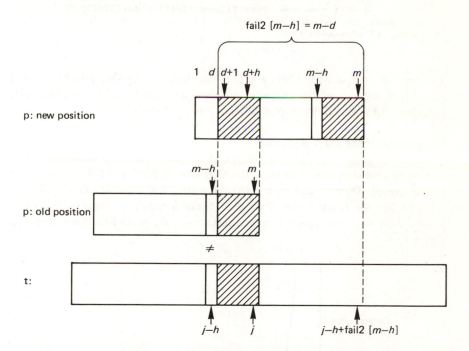

Figure 11.3   Computation of fail2 function.

As before, arrays are used to store the mappings fail1, fail2 and f. With these mappings the BM algorithm can then be implemented as follows:

```
BM algorithm:
    function search (t,p : string) : integer;
        label 99;
        var i,j : integer;
        begin
```

```
      j := m; search := 0;
      while j <= n do
        begin
          i := m;
          while (i > 0) and (index[t,j] <> index[p,i]) do
            (* i and j index into the pattern and
               the text respectively *)
            begin
              i := i-1;  (* keep on matching *)
              j := j-1
            end;
          if (i = 0) then
             begin search := i+1; goto 99 end
          else
             (* shift the pattern *)
             j := j + maximum (fail1[index[t,j]] , fail2[i])
        end;
99 : end;  (* of search *)
```

The time for computing the fail1 and fail2 functions can be shown to be $O(m+S)$ where S is the size of the character set used and the time for the BM search function is $O(n)$ (see Bibliographic notes at end of chapter).

Comparison of the Three Algorithms

The worst-case time complexities of the algorithm S, KMP and BM can be shown to be $O(mn)$, $O(m+n)$ and $O(m+n+S)$. However, the expected (average) time complexities are much better than what these figures may suggest. Empirical results show that the expected time requirements are more like those shown in figure 11.4.

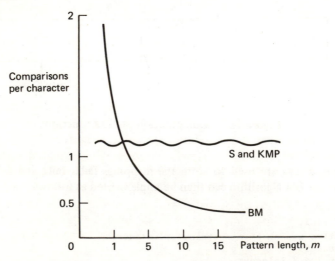

Figure 11.4 Empirical performance: number of comparisons per character *versus* the pattern length.

Algorithms S and KMP tend to be very similar in performance, with approximately 1 comparison per character. The BM algorithm, on the other hand, shows a very sharp decline as the size of the pattern increases.

Special knowledge about the text to be searched can also be used to improve the efficiency considerably. For instance, in algorithms S and KMP, it is best to search for a character in the pattern which occurs least frequently in the text, and then start searching as usual. This uses the information that different characters have different probability of occurrence. Another observation when using the BM algorithm is that fail2 does not contribute too much to the efficiency of the algorithm since repetitive substrings in a pattern are not very common. Again, empirical results support these two types of improvement. A final note is that since a simple algorithm works faster for small patterns, the search function could select the best algorithm to use based on some threshold value for the length of the pattern.

11.4 String-searching in Large Static Strings

In certain applications of string searching where the probability of finding a pattern is very small — that is, the frequency of occurrence of a pattern is low — the algorithms of section 11.3 in the best case give an almost linear behaviour (with respect to the size of the text to be searched). A technique based on hashing can be used to reduce the amount of searching dramatically. This is particularly important when the text to be searched is very large and/or where a data type string needs to be searched to find all occurrences of a pattern and the frequency of this pattern is low compared with the total size of the text. Another application would be in information retrieval where large textual documents (such as the text of this book) are to be searched to find paragraphs, sections or pages containing a given keyword.

The main idea is that if the search for a pattern is likely to fail very often, we can precede searching by a simpler and faster test to decide whether it is possible for the text to contain an occurrence of the pattern. If not, searching will no longer be necessary.

We shall describe such a fast pre-test known as *hashed 2-signature*. A text string is first partitioned into a number of *blocks* (lines, paragraphs, pages etc. of text or a fixed number of contiguous characters as a substring of the text). Each such block of text is then given a *signature* which will be a partial representation of the content of the block. A pattern then is given its own signature. The pre-test then consists of testing whether or not the signature of the pattern is 'included' in the signature of a block. If it is not, it is impossible for the pattern to occur in that block and thus the next block should be examined. If it is indeed included in the signature of the block, it is *possible* for the pattern to be in the block and thus searching is necessary.

Let us assume that a block of text is a line of characters. A signature can be a bit-vector of m bits. The signature of a line can be formed by taking every pair of adjacent characters on the line and applying a hashing function to get an integer i in the range 1 to m. Then the bit corresponding to index i in the bit-vector should be set to a '1'. Other bits in the bit-vector must be set to '0'. Now, the test for inclusion of signatures can be carried out by a simple logical operation. If S_t and S_p denote the signatures of the line (text) and pattern respectively, then if $(S_p$ and $S_t) = S_p$ then S_p is included in S_t. 'and' is the bit-wise logical **and** operation.

Let us assume $m = 24$ and the hashing function and the pre-test are defined as

```
type signature = packed array [1..m] of boolean;
var line, pattern : string;
    S_t, S_p : signature;

function hash(C1,C2 : character) : integer;
  begin
      hash := ((ord(C1) + ord(C2) mod m) + 1
  end;  (* of hash *)

function pre_test(S_t, S_p : signature) : boolean;
  (* pre_test returns true if S_t is included in S_p and false
    otherwise *)
  var i : integer;
  begin
    pre_test := true;
    for i := 1 to m do
        if S_p[i] and not S_t[i] then pre_test := false
  end;  (* of pre_test *)
```

Using the above hash function, the signature for a line of text can be computed as

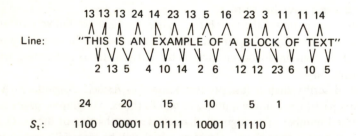

Note that, in the above, spaces are ignored since they produce a heavy bias in the hash function (see later). Now let us try a few patterns:

pattern			signature S_p				pre_test

```
22  8                24    20     15     10     5    1
 ∧  ∧
H E R E             0010  00000  00000  00100  00000  FALSE
 \ /
  V                        ↑                   ↑
  8


 4                   24    20     15     10     5    1
 ∧
A R E               0000  00000  00000  00100  01000  FALSE
 \ /
  V                                      ↑
  8


20                   24    20     15     10     5    1
 ∧
F E W               0000  10000  00100  00000  00000  FALSE
 \ /
  V                        ↑
 13


 2  1  8  18         24    20     15     10     5    1
 ∧  ∧  ∧  ∧
P A T T E R N S     0000  00110  00000  10101  00011  FALSE
 \  \  \ /
  V  V  V                  ↑↑                ↑      ↑
  6  10 17


20   16              24    20     15     10     5    1
 ∧   ∧
T O  B E            0000  10001  00000  00000  00010  FALSE
  \ /
   V                       ↑
   2


 9  4  20  18        24    20     15     10     5    1
 ∧  ∧  ∧   ∧
S E A R C H E D     0010  10100  10000  01011  01000  FALSE
 \  \  \  /
  V  V  V                  ↑     ↑ ↑    ↑     ↑ ↑
 15  6  22                 ↑


 4                   24    20     15     10     5    1
 ∧
N E W               0000  00000  00100  00000  01000  TRUE
 \ /
  V
 13


12                   24    20     15     10     5    1
 ∧
O L D               0000  00000  00010  00000  00001  FALSE
 \ /
  V                                                 ↑
  1
```

The arrows in the above diagrams indicate the '1's which are in the signature of the pattern but not in the signature of the line. Of the eight patterns, only

"NEW" has given a positive pre-test value and necessitates a complete search of the line.

This algorithm can be easily generalised to 'hashed k-signatures' where sub-strings of length k are taken and hashed accordingly. The success of this algorithm thus depends on the values of m, k and the randomness of the hashing function.

Usually m is chosen to be a multiple of the number of bits in a computer word, so that one or two machine instructions can be used to implement the pre-test function. If m is too small, signatures will be full of '1's and therefore in-effective. However, if m is too large, it may create storage overheads. A balanced signature for a block would be expected to have equal number of '1's and '0's. If a block is of length L, with a random hash function and for random strings, this implies that the size of the signatures should be approximately $1.4L$ — that is, almost a bit per character of a line (see Bibliographic notes, page 324). This in turn implies that the extra storage needed is 1.4 bits for 8 bits (each ASCII character is 8 bits long) — that is, about 18 per cent.

To ensure that the hashing function is as random as possible, any sources of bias should be removed. This is why we did not consider spaces in the above example. Other high-frequency letters (such as 'e') or bigrams (pairs of letters) could also be discarded.

The implementation of strings in the light of this 'randomised' algorithm should be modified. Obviously, this algorithm requires a signature to be assigned to each block (a partition) of a string. Every time new strings are formed by deleting, inserting or concatenating strings together, the signatures must be up-dated accordingly. This is why this algorithm is particularly suitable for *static* strings which do not change very much and therefore the update overhead is minimal. Information-retrieval applications of textual data fall into this category. If the string in question is a file of text, the signatures for each block of text can be stored in a separate file and only when a true pre-test is obtained should an actual block from the text file be examined. This is to ensure that backing memory accesses are minimised.

Exercises

11.1. Implement the ADT string using the data structure:

```
type string = packed array[1..N] of char;
```

11.2. Implement the ADT string using the augmented structure:

```
type string = record
                ch : packed array[1..N] of char;
                length : 0..N
              end;
```

11.3. How can dynamic strings be implemented where variable-length strings can be manipulated without excessive storage and without any bound on the maximum length of strings? Investigate how memory can be allocated and reclaimed as strings are created and deleted.

11.4. Find the fail function for the following strings:

> ABCACACDFA
> CONCOCT
> CONCATENATE
> ABXMNABXMNABB

11.5. Give an example of a text and a pattern where searching for the pattern using the KMP algorithm would involve comparing all the text characters twice except for the first and the last characters.

11.6. Define the ADT string as it would be appropriate for writing an editor.

11.7. Give an implementation of the ADT of exercise **11.6.** What is the performance of your solution in terms of time and space?

11.8. Re-implement the search function of exercise **11.7** by using the hashed 2-signature algorithm.

11.9. Using random patterns of varying lengths, compare the efficiency of all the string operations of your implementations to exercises **11.2** and **11.8.**

Use the following algorithm to generate random numbers in the range [0, 1].

```
var rand : real
    theta,m,seed : integer;

procedure generate;
  begin
    seed := (seed * theta) mod m;
    rand := seed/m
  end;
..........
(* main program *)
theta := 20403;
m := 16384;  (*  2**14  *)
seed := 19227;
..........
generate;
(* use rand *)
..........
```

Bibliographic Notes and Further Reading

KMP is from Knuth *et al.* (1977) and BM is from Boyer and Moore (1977). The empirical results of KMP and BM are from Horspool (1980) and Smit (1982). Horspool shows that the *s*earch for the *l*owest *f*requency *c*haracter (SLFC) and the simplified BM (SBM) are superior to algorithms S and BM respectively. Also, it is pointed out that SBM works well for computers with a search instruction. The hashed k-signature algorithm is due to Harrison (1971). He shows that the number of zeros in a signature of a string of length $L + k - 1$ is approximately $m \times e^{-L/m}$. More advanced use of strings, such as fuzzy matching, can be found in the language SNOBOL (Maurer, 1976) and in Aho *et al.* (1974).

Aho, A. V., Hopcroft, J. E. and Ullman, J. D. (1974). *The Design and Analysis of Computer Algorithms*, Addison-Wesley, Reading, Massachusetts.

Boyer, R. S. and Moore, J. S. (1977). 'A fast string search algorithm', *CACM*, Vol. 20, pp. 762-772.

Harrison, M. C. (1971). 'Implementation of the substring test by hashing', *CACM*, Vol. 14, No. 2, pp. 777-779.

Horspool, R. N. (1980). 'Practical fast searching in strings', *Software Practice and Experience*, Vol. 10, pp. 501-506.

Knuth, P. E., Morris, J. H. and Pratt, V. B. (1977). 'Fast pattern matching in strings', *SIAM Journal of Computing*, Vol. 6, pp. 323-350.

Maurer, W. D. (1976). *The Programmer's Introduction to SNOBOL*, Elsevier North-Holland, Amsterdam.

Smit, G. Dev (1982). 'A comparison of three string matching algorithms', *Software Practice and Experience*, Vol. 12, pp. 57-66.

12 'Hard' Problems and NP-completeness

So far we have studied efficient algorithms for a variety of problems. The time requirements for these algorithms varied from O(1) to O(log log n), O(log n), O(n), O(n log n), O(n^2) and O(n^3). All these algorithms are considered to be 'good' and are known as *polynomial* solutions. This is because the time requirement of each one is bounded asymptotically by a polynomial function of the size of the problem. For instance, log(n) $<$ n for all $n \geqslant 1$.

However, there are certain problems for which there exist no known polynomial algorithms. These problems are therefore considered as 'hard' problems (from a computational viewpoint). One such problem is the Travelling Salesman Problem (TSP). In TSP, a salesman is to visit n cities and the distances between each two cities are known. The problem is to find a tour of n cities with total distance less than or equal to a given value (say B) such that each city is visited exactly once (except the initial city). In abstract terms, in a weighted graph, one needs to find a simple cycle containing all vertices with total weighted length $\leqslant B$. A simple algorithm is to try all possible simple cycles. Such a solution naturally takes a maximum time of O($n!$). Such an algorithm is known as *non-polynomial* or *exponential*, since $n!$ is approximately $(n/e)^n$ and there are no integers a and N such that $(n/e)^n \leqslant n^a$ for all $n \geqslant N$. A usual variant of the TSP is when we require to find the tour with minimum cost. Let us call this problem the Optimised Travelling Salesman Problem (OTSP). Indeed, this more general problem is not computationally harder than the TSP. Since, if the general problem of minimisation (OTSP) is solved in time T, the TSP can be solved straight away in time P(T) where P(x) is a polynomial function of x. Conversely, if the initial TSP problem is solved 'efficiently', then the latter version can be solved efficiently (polynomially proportional to the former case) by applying the solution to the former version a polynomial number of times (see exercise **12.1**).

A second 'hard' problem is the Satisfiability Problem (SP). In SP, for a given logical expression with n boolean variables, it must be ascertained if a combination of n boolean values exists so that the whole expression will yield a true value (see exercise **8.5** in chapter 8). A simple algorithm for this problem is to generate all possible combinations of n boolean values and check that one satisfies the expression. This requires a maximum time of O($n^2 2^n$). Again, this algorithm is exponential. In the remainder of this chapter there are several other examples of problems like TSP and SP.

But the question arises: can polynomial algorithms be formulated to solve this type of problem? The unfortunate fact is that no proof exists either way. However, a new question then arises: given a problem for which all attempts to find an efficient polynomial algorithm have failed, can we be sure that no such algorithm exists?

To be able to classify these 'hard' problems, a hypothetical model of a computation is developed, which can then be used to identify problems of this nature.

By a *deterministic* algorithm, we mean that at any time, whatever the algorithm is doing, there is only one thing that it could do next. In effect, these algorithms capture the sequential nature of existing computers where only one operation is executed at any point of time, followed by another operation and so on.

Now, we can extend the power of a computer by providing *non-determinism*: when an algorithm is faced with a choice of several options, it has the power to *guess* the right one. Thus the non-deterministic computer has the power of evaluating an unbounded number of computational sequences in parallel. Such a computer is obviously hypothetical. It may sound absurd and unreasonable to imagine a non-deterministic computer when all real computers are deterministic and can pursue only one sequence of computation at the most. However, as we shall see, this is used as a device for deterministic computers to decide the 'hardness' of a problem.

For example, in the TSP, a non-deterministic algorithm can correctly guess the solution – that is, n arcs:

```
TSP : for i := 1 to n do
         begin
            choose an arc
            delete it from the set of arcs
         end;
```

This solution can be easily verified. Simply, all the weights of the selected arcs can be added up and it can be checked that the sum is less than or equal to B. In the satisfiability problem, a non-deterministic algorithm correctly guesses n boolean values. This solution can then be efficiently verified by checking that it satisfies the expression.

```
SP :    for i := 1 to n do
           begin
              choose a value for the ith variable (true or false)
           end
```

In these algorithms the *choose* command characterises the non-deterministic nature of the computation which can be considered to take $O(1)$ steps of non-deterministic computation. Thus in the TSP, the guessing phase takes $O(n)$ and the verification stage also takes $O(n)$. Therefore the total time is $O(n)$. Similarly, the SP needs a time of $O(n)$ for the guessing phase and $O(n^2)$ to verify the answer, the total time being $O(n^2)$. These two algorithms are therefore polynomial on a non-deterministic computer.

The above terminology is meant for *decision* problems – that is, problems whose solutions are either 'yes' or 'no'. However, general non-decision problems are not any harder. These include *search* problems (that is, problems which require a particular solution with a special property, such as to minimise or maximise a function) and *enumeration* problems (that is, problems which need to find how many solutions satisfy a property). The OTSP is a search problem. If the TSP requires all the tours with cost $\leqslant B$, then it would be an enumeration problem. Search problems can be solved in time T where T is proportional to a polynomial function of T', where T' is the time for their corresponding decision problems (see above for TSP and OTSP). The complexity of enumeration problems is usually determined by the size of their input and output sequences. For instance, in the enumeration version of TSP, we can construct instances such that the total number of tours with cost $\leqslant B$ is exponential (in terms of n). In these cases, almost all the time is spent on describing the output. For this reason we usually exclude enumeration problems from our discussions and concentrate on decision and search problems.

Now we can broadly classify problems as follows:

(1) *Undecidable problems* (unsolvable problems) for which there are no algorithms. For instance, deciding whether a general program halts or not is undecidable.

(2) *Intractable problems* (provably difficult problems) with no polynomial algorithms to solve them. Only exponential algorithms can be expected to solve these. For example, deciding the emptiness-of-complement problems for extended regular expressions (formal languages) is intractable.

These two classes of problems are outside the scope of this book. For further details, see the bibliographic notes at the end of the chapter.

(3) *NP problems* (that is, *N*on-deterministic *P*olynomial (NP) problems) are those for which there exist *polynomial* algorithms on a non-deterministic computer (see below).

(4) *P problems* (that is, *P*olynomial (P) problems) are those for which there exist *polynomial* algorithms on a *deterministic* computer. All the problems discussed in this book so far are in P.

12.1 NP Problems and NP-complete Problems

The TSP and SP are both in NP since, given a guessed solution, it takes $O(n)$ and $O(n^2)$ to verify the solutions respectively. Obviously all the problems in P are also in NP. Simply, guessing power is not used for problems in P. Therefore P is a subset of NP (figure 12.1).

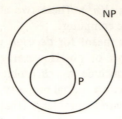

Figure 12.1   P is a subset of NP.

We mentioned that for SP and TSP, it has not yet been proved whether they are in P or not. In other words, no-one has yet proved if P $\neq$ NP or N = NP. There are many more NP problems for which researchers have failed to find efficient solutions in P. Because of this, it is safely conjectured that P $\neq$ NP. Until such a proof is found, we need somehow to isolate problems in NP, such as SP and TSP, that seem to be 'harder' than other NP problems.

NP-complete Problems

There is a class of NP problems which is, in a sense the *hardest*. These are named *NP-complete* (NPC problems). For a problem X to be in NPC, firstly, it must be in NP. Then we should be able to *transform* any other problem in NP by a *simple transformation* to an instance of problem X.

   In effect, if this is done, then if a good (polynomial, deterministic) solution is found for X, then all problems in NP can be solved efficiently. In other words, if X is NP-complete and belongs to P — that is, a good (polynomial, deterministic) solution for X is found — then P = NP. This is because the transformations are *simple* (that is, they take only a polynomial amount of time). Also, if for a NP problem Y, X can be transformed to Y, then Y is also a NP-complete problem. The former property, therefore, states that NP-complete problems are the hardest problems since if any NP-complete problem is solved in polynomial time, then all problems in NP are solvable in polynomial time. Conversely, if P $\neq$ NP, then no NP-complete problem can be solved polynomially. As a consequence, if one NP problem is proved to be 'intractable', then all NP-complete problems are intractable. This is another reason why NP-complete problems are thought to be the hardest.

   As an example of a simple polynomial transformation, let us consider the Hamiltonian Cycle Problem (HCP). A Hamiltonian cycle is a *simple* cycle of a graph G containing every vertex of V. The HCP is then to decide whether a given graph G has a Hamiltonian cycle. One way to transform an example of this problem to an example of the TSP problem is as follows:

1. Assign a weight of 1 to each arc of the graph G.

2. Find a TSP tour of cost $\leqslant n$ (assuming that G has n vertices).

The complexity of this transformation of labelling the arcs of the graph G is obviously polynomial (in terms of e, the number of arcs) and from a complexity point of view it is regarded as 'simple'. This transformation suggest that if TSP is solved efficiently, then HCP can be solved efficiently. Furthermore if HCP is NP-complete, TSP must also be NP-complete.

The easiest way to prove that a problem is NP-complete (that is, it is *very hard*) is to find a simple method to transform a problem already known to be in NPC to the problem in question. But the difficulty is that we must first find at least one problem in NPC. Fortunately Stephen Cook found just one such problem in 1971. He proved that the satisfiability problem (SP) is in NPC. The proof is very complex and difficult. Broadly speaking, he used a formal model of computation (a Non-Deterministic Machine) and proved that for any given problem for this machine to solve, there exists an instance of the SP problem.

The TSP problem is proved to be in NPC by transforming the SP problem into TSP. This can be done in stages. One approach is to transform SP to K-clique, then to Vertex cover, then to Hamiltonian cycle and finally to TSP. Each transformation should be polynomial and hence the composition of a finite number of them should be polynomial.

$$SP \rightarrow K\text{-clique} \rightarrow \text{Vertex cover} \rightarrow \text{HCP} \rightarrow \text{TSP}$$

where

K-clique: Does a graph have a clique of size K — that is, a complete sub-graph (every pair of distinct vertices in the sub-graph is connected by an arc) of G with K vertices.

Vertex cover: Does a graph have a vertex cover of size k — that is, a set S of k vertices such that each arc of G is incident upon some vertex in S.

(see exercise **12.4**.)

Although the problems stated here may sound artificial and academic, many real-world problems have proved to be NP-complete, among these being:

- To pack files on disc storage to minimise the number of disc packs
- Scheduling independent tasks optimally on multiple processors using non-preemptive schedules

etc.

So far, more than 300 problems in computer science, mathematics, operational research, number theory, game theory etc. have proved to be NP-complete. Figure 12.2 shows the conjectured relationship between P, NP and NP-complete problems.

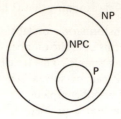

Figure 12.2   Classification of NP problems.

12.2   Methods for Solving 'Hard' Problems

Even if a problem is in NPC, from a practical point of view we cannot abandon it and we often need to find some 'good' solutions. Naturally, problems in NPC, in their most general form, always require exponential algorithms for their solution. However, using the TSP as a typical example, we will now discuss a few techniques for solving such 'hard' problems in a 'reasonable' amount of time.

12.2.1   Exhaustive Search

In an exhaustive search, all possible configurations of a problem are generated and evaluated, and then the appropriate one is selected (to optimise certain functions). For the TSP, this requires the generation of $n!$ different cases. So, this approach is viable only for small problems (when n, the number of cities, is small).

Note that the dynamic programming approach to algorithm design is a form of exhaustive search. Although it may reduce the time requirement, it cannot reduce the exponential search space of a NP-complete problem to a polynomial one. For the TSP, a dynamic programming solution will reduce the time complexity from $O(n!)$ to $O(n^2 2^n)$ (see exercise **12.2**).

12.2.2   Semi-exhaustive Search using Special Knowledge about the Problem

When doing an exhaustive search, by using special properties of the problem, many of the paths need not be searched. A technique known as *Branch and bound* is one approach to reduce the time requirements of such problems. However, it must be noted that the worst-case time complexities still remain exponential.

Branch and Bound

For many search problems such as the TSP, as we try to generate and test all possible tours (using backtracking), we may be able to put a *bound* on certain partial solutions and hence exclude them from further search. (For enumeration problems we only prune the branches which cannot possibly lead to solutions. So no bound function is required. An example is the 8-queen problem. See exercise **12.6**. This strategy is known as *Branch and exclude*.)

In branch and bound, depending on how 'tight' the bound function is, better algorithms can be derived. Let us take the TSP problem of figure 12.3. We can solve the problem in the following manner: start with node E (say) and find the value of the *bound* function for this node (v, say); that is, all solutions including node E have cost $\geq v$. Now, the next node in the route has 4 possibilities among the remaining nodes in the graph — that is, A, B, C or D. For each such 'partial' solution we calculate a new bound value. Initially we set the cost of the 'best tour found so far' to a large number. Now, any partial solution with bound value greater than the 'best tour so far' can be excluded from further expansion. As the problem is expanded to generate all possible cases, when a complete solution is found, if its cost is less than the best solution found so far, we replace the 'best tour so far' to denote this solution.

We choose to expand and process the partial solutions in non-decreasing order of their bound values, since we wish to get the smallest cost without having to generate numerous useless configurations of the problem (see below).

The crucial part of the branch and bound algorithms is determining a good bound function F. If F is too simplistic, probably very few branches may be excluded. For instance, for the TSP if the partial tour is

$$v_1, v_2, v_3, \ldots, v_m$$

a simple bound function F might then be

$$F = \sum_{i=1}^{m-1} \text{weight of the arc } (v_i, v_{i+1})$$

That is, after visiting town v_m, the salesman goes straight back to his base (town v_1). However, a 'tighter' lower bound is much preferred. One such good bound function can be found by first observing that

$$\text{Cost of a complete tour} = 1/2 \sum_{i=1}^{n} \text{arc-}i_1 + \text{arc-}i_2$$

where arc-i_1 and arc-i_2 are the weights of the arcs adjacent to node i in the tour.

$$M \equiv \text{Minimal cost of a tour} \geq 1/2 \sum_{i=1}^{n} \text{arc-}i_1 + \text{arc-}i_2$$

where arc-i_1 and arc-i_2 are the two arcs adjacent to the node i with minimum costs.

Therefore, we can choose $F \equiv M$ for each partial solution as a bound function, with the exception that those arcs which have already been chosen in the search as part of the solution take precedence over the 'two minimum' arcs as suggested above. Let us use this bound function together with this technique to find a minimal tour of the graph of figure 12.3. Figure 12.4 shows how the solution is derived by constructing a tree of partial solutions.

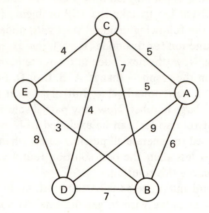

Figure 12.3   Example of the TSP problem.

Starting at node E

$$\text{Min cost} \geqslant 1/2\{(3+4) + (4+4) + (5+5) + (3+6) + (4+7)\}$$
$$\geqslant 22.5$$

Then, there are four choices for the next node in the tour — A, B, C or D. For each partial solution, (E,A), (E,B), (E,C) and (E,D), a new bound should be calculated. For instance, for node D, the lower bound is $1/2 \{(8+3) + (4+4) + (5+5) + (3+6) + (4+8)\} = 25$. This is repeated for other nodes, giving bounds 22.5, 23 and 22.5 for nodes C, A and B respectively. From these 4 nodes, the one with the smallest lower bound is chosen and expanded. Nodes C and B have bounds 22.5. Let us choose node C to be expanded next. Now, the whole process is repeated for this node. When a complete tour is found, the 'best tour so far' is accordingly modified. For instance, when the tour (E,C,D,B,A,E) is found with cost 26, this replaces the old minimal tour which initially was set to a large number. When the bound for a node exceeds this 'best tour so far', it should be pruned and therefore it will not be expanded for searching. These nodes are indicated by a × in figure 12.4. The bold numbers **1** and **2** in figure 12.4 indicate

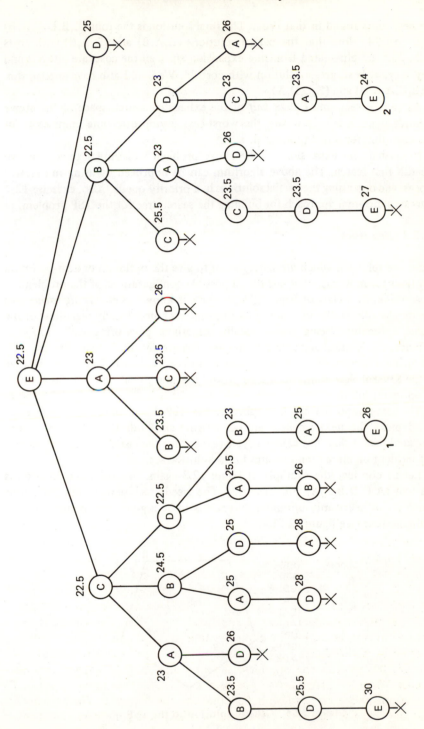

Figure 12.4   Branch and bound solution of the TSP.

the best tours found in that order. The final solution is the tour (E,B,D,C,A,E) with cost 24. Note that the partial solutions (E,A,B) and (E,A,C) with costs 23.5, 23.5 are eliminated from the expansion since all the costs are integers and they may, at best, give a solution with cost 24. We could also have applied this to (E,C,A,B,D) and (E,C,D,A).

As can be seen, more than *half* of the exhaustive search space of the above example is excluded. However, the worst-case asymptotic time complexity of this algorithm remains exponential.

Note that the expansion of the above problem was similar to a recursive breadth first search. The above algorithm can be improved by using an iterative process and organising the partial solutions in a priority queue. Also, exercise **12.7** suggests a different approach for building the search tree for the TSP problem.

12.2.3 Heuristics

These are solutions which *intuitively* seem to give the optimum or near-optimum solutions for most cases *but not* for all possible configurations of the problem.

A well-known class of heuristic algorithms is known as *greedy algorithms*. In greedy algorithms, the idea is to make optimum choices locally regardless of the global optimality. Being greedy locally sometimes pays off globally too. For instance, the Kruskal algorithm of chapter 7 is greedy. Each time, the arc with the smallest weight which does not create a cycle is chosen. Indeed, in the case of the Kruskal algorithm, it could be proved that the global solution was actually an optimum solution.

The same approach can be applied to the TSP. Each time, the *choose* command picks the next smallest arc (in weight) and adds that to the solution, *provided that* it does not create a cycle (except for the last arc) and each node is not incident on more than two arcs in the solution set.

Let us consider the example of figure 12.3 again. One optimum solution is the tour (A,C,D,B,E,A) with cost = 24. The greedy technique would give a tour which is not necessarily optimum, but it would be a good approximation to an optimum tour (see figure 12.5).

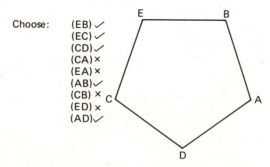

Figure 12.5   Greedy solution to the TSP.

The greedy solution is the tour (A, D, C, E, B, A) with cost = 26. This greedy solution is about 8 per cent more costly than an optimal tour. One aspect of heuristic algorithms is to calculate how close the derived solutions are to the optimum solutions. This is, however, outside the scope of this book.

12.2.4 Optimisation

In this approach, we usually pick a solution (randomly, say), and then we apply some local transformation to the given solution to improve it. Naturally, the transformations should be very simple and not too costly. The process can be repeated as many times as required.

As an example, for the TSP, a simple transformation known as *2-opting* can be used to find better tours. (This can be generalised to *k*-opting.) Let us take the tour of figure 12.6 as an initial solution to the TSP.

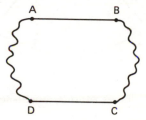

Figure 12.6   Initial solution of TSP including arcs AB and CD.

A new tour can be constructed from the tour of figure 12.6 by removing the arcs (A,B) and (C,D) and replacing them with arcs (A,C) and (B,D) (figure 12.7). If the cost of the new solution is less than the cost of the initial solution, we will adopt the new solution.

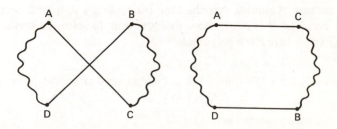

Figure 12.7   New tour resulting from a simple transformation.

Figure 12.8 shows how the optimum solution is derived from the initial non-optimum tour.

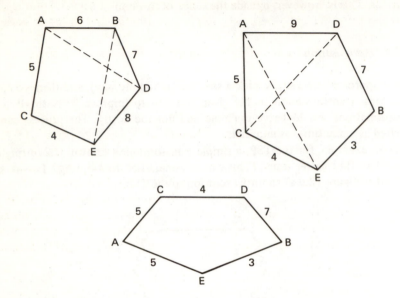

Figure 12.8   The sequence of transformations to derive the optimum solution.

In practice, it is often too costly and time-consuming to allow all possible transformations to be applied to the problem. Therefore, as soon as an acceptable solution is reached or we run out of processing time, the optimisation can be stopped (see also exercise **12.9**).

Exercises

12.1. The 'search' problem version of the TSP is to find the tour with minimal total cost. Assuming that the time for finding a tour with cost less than or equal to B is $F(n)$, show that the time to solve the minimal tour is $P(F(n))$ where P is a polynomial in F.
Hint:

Min cost for a TSP tour $\leqslant 1/2 \Sigma$ Two minimal arcs incident on each vertex
$$\leqslant m \text{ (say)}$$

Use a binary search operation:

If there exists a tour with cost $\leqslant m/2$ **then** repeat for cost $= m/4$
else repeat for cost $= 3m/4$, etc.

12.2. Can dynamic programming be used for solving the TSP? Try constructing a solution by first finding minimal tours of size 1, then of size 2, and so on. Find the time complexity of this algorithm.

12.3. Can divide and conquer be used as a heuristic or to find an optimal solution for the TSP when the graph is that of cities and distances between them?
Hint: Consider a rectangle containing all the cities, then subdivide the rectangle.

12 4. Construct a simple transformation from the k-clique problem to the vertex cover problem.
Hint: G = (V, E). Construct G' = (V, E')
where E' = complement of E; that is, $\{(v, w) \mid v \neq w$ and $(v, w) \in E\}$.
If S $\subset$ V is a k-clique, then V$-$S is a vertex cover for G'.

12.5. Find the optimal tour of the graph of figure 12.3 using branch and bound by using a priority queue to hold the partial solutions.

12.6. Design and implement a program for finding all the solutions to the 8-queen problem using the branch and exclude technique. Note that the tree of the problem should be generated systematically and dynamically on demand. For simplicity, first try to tackle the problem of finding only one solution.

12.7. A different approach to decomposing the TSP problem is first to consider all the tours. Then we split these into tours that include arc e1 and tours that exclude arc e1. Next, for each of these two categories, we further split the tours into those that include and that exclude another arc e2, and so on.

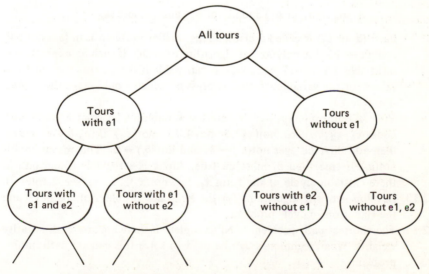

Figure 12.9 Decomposition of the TSP problem

Using the same bound function as discussed before, develop an algorithm to find the minimal tour.

12.8. *Könizberg park problem (Euler's problem)*

In a park with a number of bridges, find a path which passes over each bridge exactly once. Is this problem NP-complete? If not, find a polynomial algorithm to solve it.

Hint: When represented as a graph, there is a solution if each node has an even number of arcs incident upon it.

12.9. It is required to place a fixed number of integrated circuit (IC) components on a printed circuit board (PCB) such that the total cost of the wires is minimal. Given an initial placement of the IC components, find a simple transformation which would reduce the total cost by exchanging ICs. Assume that the cost is given in terms of total length of wires and the length of a wire is the Euclidean distance between two ICs.

12.10. Generalise the 2-opting technique to k-opting where $k \geqslant 2$.

12.11. The *assignment problem* is stated as: given n people and n jobs, assign each person to a job in an optimal fashion — that is, if a_{ij} is the cost of assigning person i to job j, then minimise the cost

$$\sum_i a_{ip_i}$$

where p is a permutation of numbers $1 . . n$.

How does this problem relate to the TSP problem? Try to find some good solutions for the assignment problem.

12.12. A subset of the satisfiability problem known as *2-satisfiability (2-sat)* is defined for logical expressions of the form:

Alpha $\equiv (a_1$ or $b_1)$ and $(a_2$ or $b_2)$ and $\ldots$ and $(a_n$ or $b_n)$.

Each $(a_i$ or $b_i)$ is called a clause. This smaller problem is in fact not NP-complete. Find a polynomial algorithm to decide if such an expression is satisfiable. (A logical expression is satisfiable if there exists a set of truth values for the variables in that expression which would make the expression true.)

Hint: Assume $x_1, x_2, \ldots, x_n$ are the variables. Construct a graph with nodes $x_1, x_2, \ldots, x_n$, not(x_1), not(x_2), $\ldots$. If (a_i, b_i) is a clause then add directed eges not(a_i) $\rightarrow b_i$ and not(b_i) $\rightarrow a_i$. This means that if not(a_i) is true then b_i must be true. Alpha is satisfiable if and only if there exists no cycle of the form $x_i, \ldots,$ not(x_i), $\ldots, x_i$ (see chapter 7).

12.13. Develop heuristic algorithms for the knapsack problem of chapter 8 and the Hamiltonian cycle problem.

12.14. The shortest path problem is NP-complete if there are arcs with negative weights. What techniques can be used to solve this general path-finding problem?

12.15. The shortest path problem is not NP-complete when there are arcs with negative weights, but there are no cycles of negative length. How can the techniques of chapter 8 be used to design an algorithm for this problem? What would its time complexity be?

12.16. The longest path problem is in fact NP-complete. Verify, by examples, that this problem is very 'hard'.

12.17. Is the longest path problem still very 'hard' if there exist no directed cycles in the digraph? (This problem is usually referred to as the Critical path problem. It is used in project management where each task can be followed by another task after a prescribed time lapse. The objective is to find the maximum duration of the entire project.)

12.18. Partition problem

Set $A = \{a_1, a_2, \ldots, a_n\}$ where $B = \sum\limits_{1 \leqslant i \leqslant n} a_i$. Find A', a subset of A,

such that

$$B/2 = \sum_{a_i \in A'} a_i = \sum_{a_i \in A - A'} a_i$$

Use dynamic programming to design a polynomial algorithm for the partition problem.

12.19. *Bin-packing problem*

There are a number of bins each of capacity C and n items of sizes $S_1, S_2, \ldots, S_n$. Pack the n items in M bins such that the total size of items in each bin does not exceed C. Find the minimal M — that is, the least number of bins to accommodate the n items.

12.20. Find a heuristic algorithm for the bin-packing problem.

12.21. *Colouring problem*

Relation R is defined on the set S. Find a minimal k such that there exists a surjective function $f: S \rightarrow \{1, 2, 3, \ldots, k\}$ such that if $(a, b) \in R$ then $f(a) \neq f(b)$.

Hint: find the maximal complete sub-graph. The number of nodes in this clique is the minimum k which is required.

Bibliographic Notes and Further Reading

For reasons of brevity and space, the treatment of NP-completeness has been very informal and incomplete. For more detailed treatment of NP-completeness see Aho *et al.* (1974) and Garey and Johnson (1979). Techniques for transforming problems and other theoretical issues can be found in Garey and Johnson (1979). Hu (1972) has detailed analyses of the methods described in this chapter. In particular, he discusses cases when the greedy technique gives the optimal solution. The solution to exercise **12.9** and similar problems in computer-aided design can be found in Breuer (1972). Cook's theorem is proved in Aho *et al.* (1974) and also in Garey and Johnson (1979). Possible solutions for the assignment problem can be found in Munkers (1957) and Kurtzberg (1962). The TSP problem for large graphs is discussed in Litke (1984). A historical perspective of complexity theory and the solution to exercise **12.3** can be found in Karp (1986).

Finally, a note on the knapsack problem: this problem is said to NP-complete despite the fact that we have found an $O(mC)$ algorithm for it. The reason is that in complexity theory each problem is encoded as a string of symbols using a 'reasonable encoding scheme'. The size of this string is then the size of the problem. The integers in this encoding scheme are represented in binary. Thus the number C in the knapsack problem is represented as a binary number with $L=\log C$ (base 2) bits. Thus the time complexity of this algorithm is $O(mC) = O(m2^L)$. This is obviously an exponential algorithm. However, in practice we can usually put a maximum bound on the length L (for example, 16 in a 16-bit computer) and get reasonably efficient algorithms. These algorithms are known as *pseudo-polynomial* algorithms. To distinguish these types of problems from others, a problem is said to be *strongly NP-complete* if it remains NP-complete even if numbers are represented in unary. The knapsack and the partition problems are not strongly NP-complete and have pseudo-polynomial solutions. However, SP, TSP and the clique problem are strongly NP-complete. (More on this topic can be found in the book by Garey and Johnson (1979)).

Aho, A. V., Hopcroft, J. E. and Ullman, J. D. (1974). *The Design and Analysis of Computer Algorithms*, Addison-Wesley, Reading, Massachusetts.

Breuer, M. A. (1972). *Design Automation of Digital Systems. Volume 1: Theory and Techniques*, Prentice-Hall, Englewood Cliffs, New Jersey.

Garey, M. and Johnson, D. (1979). *Computers and Intractability: A Guide to the Theory of NP-Completeness*, Freeman, San Francisco.

Hu, T. C. (1982). *Combinatorial Algorithms*, Addison-Wesley, Reading, Massachusetts.

Karp R. M. (1986). 'Combinatorics, complexity and randomness', *CACM*, Vol. 29, No. 2, pp. 98–110.

Kurtzberg, J. M. (1962). 'On approximation methods for the assignment problem', *Journal of ACM*, Vol. 9, No. 4, pp. 419–439.

Litke, J. D. (1984). 'An improved solution to the travelling salesman problem with thousands of nodes', *CACM*, Vol. 27, No. 12, pp. 1227–1236.

Munkers, J. (1957). 'Algorithms for the assignment and transportation problems', *Journal of SIAM*, Vol. 5, pp. 32–38.

Solutions to Selected Exercises

Chapter 1

1.2. Assume that: probability(C being in the array) = 1/2 and
probability(C not in the array) = 1/2.
The worst and best cases remain the same, but the average case should be
calculated as

$$1/2n \sum_{i=1}^{n} T_i + 1/2\, T_n$$

1.4. No. For details see exercise **2.5** of chapter 2.

1.5. The loop structure can be modified to

```
while A[i] <> C do ....
```

1.7. In the context of searching ordered arrays:

```
Search(C, L, U) : (* search for C in the sequence A[L]..A[U] *)
     ratio := (C - A[L]) / (A[U] - A[L]);
     index := trunc(ratio * (U-L) + 0.5) + L;
     if (index < L) or (index > U) then failure;
     element := A[index];
     if C = element then success
     else if C > element then search(C, index+1, U)
          else search(C, L, index-1);
```

1.8. Bad solution:

```
              function evaluate(x : real) : real;
                begin
                   evaluate := 2*x + 4*x*x + 5*x*x*x*x + 13*x*x*x*x*x
                end;
```

Good solution:

```
              function evaluate(x : real) : real;
                begin
                   evaluate := x*(2 + x*(4 + x*(5 + 13*x)))
                end;
```

1.12.

```
function e(n : real) : real;        function e(n : real) : real;
  var i : integer;                    var i : integer;
      t : real;                       begin
```

341

```
begin                          e := 0;
  t := n;                      for i := 1 to 10 do
  e := n;                          e := e + exp(n,i)/fact(i)
  for i := 2 to 10 do          end;
    begin                      where exp and fact are functions
      t := t*n/i;              returning n^i and i! respectively.
      e := e + t
    end
end;
```

1.13. $3n^2 + 10n = O(n^2)$

   $1/10\, n^2 + 2^n = O(2^n)$

   $21 + 1/n = O(1)$

   $\log n^3 = O(\log n)$

   $10 \log 3^n = O(n)$

   $(f(n) + 10)^2 = O(n^2)$

Chapter 2

2.3.   (i)

```
for i := 1 to n do S;  ≡   procedure A(i : integer);
                         ` begin
                             if i <= n then begin S; A(i+1) end
                           end;
                           ......
                           A(1);
```

   (ii)

```
while C do S;  ≡           procedure A;
                             begin
                               if C then begin S; A end
                             end;
                             ......
                           A;
```

   (iii)

```
repeat S until C;  ≡       procedure A;
                             begin
                               S;
                               if not C then A
                             end;
                             .....
                           A;
```

   (iv) For representing recursion using loop structures, see the final part of section 3.1.

2.4.

```
procedure for_all (X : zet; procedure p(x : xtype));
  begin
    for all elements x in the set X, execute p(x);
  end;
......
for all x in X do S(x) ≡ for_all(X, S);
```

2.5.

```
search(x)    : pre: true
               post: if {y | A[y] = x} <> {} then
                        search = minimum ({y | A[y] = x})
                     else search = 0.

b_search(x) : pre: true
               post: if ∃ i A[i] = x then search = i
                     else search = 0
```

In other words, any index i containing x may be returned, depending on how the binary search proceeds.

2.12.

```
let elems ≡ (a₁, a₂, ... `, aᵢ)  0 <= i <= s. When i = 0, elems = {}.
initialise(s) : pre :   true
                  post: empty(s)
push(s,a)     : pre : not full(s)
                  post: elems'= (a₁, a₂, ... , aᵢ, a)
pop(s)        : pre : not empty(s)
                  post: elems' = (a₁, a₂, ..., aᵢ₋₁)
top(s)        : pre : not empty(s)
                  post: top' = aᵢ
empty(s)      : empty' = (i = 0)
full(s)       : full'  = (i = n)
```

Chapter 3

3.1. Let us assume that the sequence of characters is terminated by a '.':

```
procedure reverse;
  var ch : char;
  begin
  read(ch);
  if ch <> '.' then reverse;
  write(ch)
end;
```

3.2. After the second initialise(S) operation is performed, if stack S was not empty, then the elements of the old stack will be retained in the memory despite the fact that we cannot access them any longer. If used frequently, the program will soon run out of memory. To remedy this, a new operation 'clear' is needed which would dispose of all the elements of the old stack.

3.3.   Let us try the ADT STACK:

```
type STACK : LIST;

procedure initialise_stk(var S : STACK);
  begin
    initialise_lst(S)
  end;

procedure push(var S : STACK; a : elemtype);
  begin
    insert(S, a, first(S))
  end;

procedure pop(var S : STACK);
  begin
    if empty_lst(S) then error
    else
      delete(S,first(S)
  end;

function top(S : STACK) : elemtype;
  begin
    if empty_lst(S) then error
    else
      top := retrieve(first(S), S)
  end;

function empty_stk(S : STACK) : elemtype;
  begin
    empty_stk := empty_lst(S)
  end;
```

3.5.   First, we can remove the tail recursion:

```
  procedure hanoi(n,i,j,k : integer);
    label 100;
    var t : integer;
    begin
100 : if n = 1 then (* move the top disc from pole i to pole k *)
      else begin
              hanoi(n-1,i,k,j);
              (* move the top disc from pole i to pole k *)
              (* hanoi(n-1,j,i,k) *)
              n := n-1;
              t := i; i := j; j := t;
              goto 100
            end
    end;
```

Now we use a stack to transform the only recursive call. The return address from hanoi is either the main program or inside procedure hanoi immediately after the recursive call. Let us denote these by −1 and 0 respectively. Thus

```
program hanoiprog(input, output);
   type stack = ; (* the ADT STACK *)
   var n : integer;

   procedure hanoi(n,i,j,k : integer);
      label 100, 200;
      const r = -1;
         (* r represents the call from inside the procedure *)
      var mode, t : integer;
      begin
100 : if n = 1 then
            begin
               (* move the top disc from pole i to pole k *)
               write('MOVE FROM/TO ');
               writeln(i:3,k:3)
            end
         else
            begin
               (* hanoi(n-1,i,k,j); *)
               push(S,n); push(S,i); push(S,j);
               push(S,k); push(S,r);
               n := n-1; t := k; k := j; j := t;
               goto 100;
      200 :
               (* move the top disc from pole i to pole k *)
               write('MOVE FROM/TO '); writeln(i:3,k:3);
               (* hanoi(n-1,j,i,k);   TAIL RECURSION  *)
               n := n-1; t := i; i := j; j := t;
               goto 100
            end;
         if not empty(S) then
            begin
               mode := top(S); pop(S);
               k := top(S); pop(S);
               j := top(S); pop(S);
               i := top(S); pop(S);
               n := top(S); pop(S);
               if mode = r then goto 200
            end;
      end; (* of hanoi *)

begin (* of hanoiprog *)
   writeln('Type the number of discs');
   readln(n);
   initialise(S);
   push(S,n); push(S,i); push(S,j); push(S,k); push(S,0);
   (*  0 indicates the call from the program *)
   hanoi(n, 1, 2, 3)
end.
```

3.12. The clue lies in the fact that L is a value parameter of the procedure X.

3.14. When evaluating an expression, each time an identifier is evaluated, its result in Reverse Polish can be stored in the table so that it may not be re-evaluated again. For instance, in the following case:

$$A = B + 10/C$$
$$C = 25 * D + E - 2 * B$$
$$D = B/18$$
$$E = D - 25 + B$$
$$B = 10 * 123 - 10768/43$$
$$10 * A$$

B will be evaluated 5 times. These re-evaluations should be avoided.

3.16. *Hint:* assuming that numbers are positive integers, then

$$A + B: \text{if } (MAXINT - B) < A \text{ then error}$$
$$A - B: -$$
$$A * B: \text{if } (MAXINT/B) < A \text{ then error}$$
$$A / B: -$$

3.17. The roundabout can be divided into a number of segments. Every road has a queue of cars waiting to enter the roundabout and exit at a required road. Each segment is also a queue of cars in a segment of the roundabout.

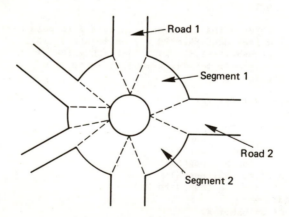

```
Type NoOfRoads = 1..5
ADT road : identified by integers 1, 2, 3, ...
ADT car : Model : composed object
     Operations : iden(car) : CarIden; (* identification of a car *)
                  turn(car) : road;    (* the exit road of a car  *)
                  create : car
                  etc.
```

```
ADT queue of car
ADT segment : queue of car
Type roundabout_section : record
                            case kind : (road_sec, segment_sec) of
                              road_sec : (rd : road;
                                           q : queue);
                              segment_sec : (s : segment)
                            end;

ADT RING : Circular LIST of roundabout_section
          Operations : initialise
                       next
                       previous: These two return the next and previous
                                 roundabout_section of a RING respectively.
                       insert
                       delete
                       retrieve(ring, position) : roundabout_section
var roundabout : RING
    p : position; (* into the ring roundabout *)
    c : car;
    rs : roundabout_section;
    s : segment;
```

Let us assume that c is a car about to enter the roundabout, that is

> rs = retrieve(roundabout, p)
> rs.kind = road_segment
> c = front(rs.q)

Now the next segment of the roundabout should be retrieved. If it has some room, then the car c should be deleted from the front(rs.q) and be added to the segment queue of the next section of the roundabout and the position p should be advanced to this segment.

Now p should be advanced in the ring until a road section is found which is the same as turn(c). In this case, c can be removed from the segment section (by a dequeue operation).

Chapter 4

4.1. pre-order: abcdefghjkilmnpq
in-order: cbdfegajhknmpqli
post-order: cdfgebjkhnpqmlia

4.2.

```
procedure in_order(n : node);
  var t : node;
  begin
    if not null_node(n) then
      begin
```

```
      write('(');
      t := leftmost_child(n);
      in_order(t);
      write(retrieve(n));
      t := right_sibling(t);
      in_order(t);
      writeln(')')
   end
end;
```

4.6.   The parent node of a node n can be found by either

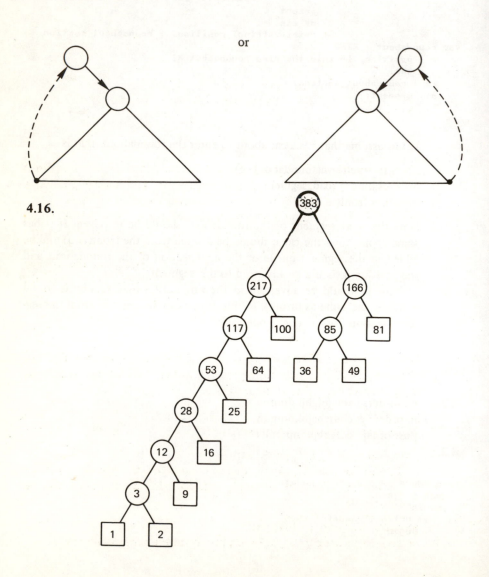

or

4.16.

The average code length is

$$[(1+2)*7 + 9*6 + 16*5 + 25*4 + (64+36+49)*3 + (100+81)*2] / 383 = 2.8($$

4.21. e = the number of edges in a tree

$e = 2*n$ since each internal node has two edges to its children.

Also, each node except the root has an edge to its parent node, thus

$e = n + x - 1$ where x is the number of external nodes.

Therefore $n + x - 1 = 2 * n \Rightarrow x = n + 1$. *Q.E.D.*

4.27.

```
function cons(n : integer) : tree;
  var i : integer; tL,tR : tree;
  begin
    if n > 0 then
      begin
        tL := cons(n div 2);
        read(i);
        tR := cons(n - (n div 2) - 1);
        cons_tree(i, tL,tR)  (* construct a binary tree *)
      end
  end;
```

4.28 *Hint:* use the function cons of exercise **4.27** above, but modify the read(i) operation to 'i := retrieve(L,p); p := next(p,L);'. The position p must be initialised to first(L) prior to calling the above procedure.

4.29. A straightforward solution is to write a procedure get_next which would return the next element of the tree in the in-order traversal of that tree (this needs an explicit stack). Then the procedure cons of exercise **4.27** can be modified by replacing 'read(i)' to 'get_next(i)'. (Note that for the first element, the element as far to the left of the tree should be returned.)

Chapter 5

5.5. Define *expressions* as a set of pairs (id, expr) where id is an identifier and expr is a list of symbols defining that identifier. This set can be defined as an abstract data type with operations:

procedure clear;
(* To build an empty set expressions *)

procedure add(id : IdType; e : expr);
(* To add the pair (id, e) into the set expressions. If a pair (id, e') exists in the set, or if e is an empty list, the addition operation will be aborted *)

procedure remove(id : IdType);
(* To remove the pair (id, e) (if any) from the set expressions *)

procedure GetExp(id : IdType; **var** e : exp);
(∗ To retrieve the definition of an identifier. In other words, if the pair (id, exp) exists in the table, the parameter e will return exp as its value; otherwise an empty list will be returned ∗)

5.10. If the data type of S1 and S2 is an array or a record (containing an array), then this is O.K. However, if this type is a pointer or an index into an array, then, S2 := S1 simply assigns a reference to the set S1 to S2. The actual structure of S1 is not therefore copied. This means that the actual structure of S1 will be shared with that of set S2.

Chapter 6

6.3. Address: 0 1   2   3 4   5   6   7 8 9 10 11 12 13 14 15
      Content: 16 1 49 64 4 36 81 100   9 25 121

6.4. This is because 4 and 16 are not co-prime.

6.13. An outline of the program would be:

```
initialise(trie);        (* to create an empty trie *)
while there are more words do
  begin
    getword(word);        (* to read a word form the input *)
    insert(word, trie)    (* add word to trie *)
  end;
print(trie);
```

The necessary data structures can be a set of nodes organised as a tree where each node has 2 pointers (left_child, right_sibling), a count and a character. To satisfy the requirements of the assignment regarding the count of complete words and counts of proper prefixes, we can use the count field of each $ node to represent the frequency of complete words and the count field of non-$ nodes would assume that every word has only been inserted once. Therefore, when a new word is being added to the trie, if it is found that the word was already in the trie, the count fields of the nodes on the path to the root should be decremented.

An alternative to this structure would be an array [1. .27] of pointers to represent the chain of the right-sibling nodes, but it would be a complete waste of storage.

For printing the trie, a traversal similar to pre-order traversal can be used. As we get deeper in the trie level by level, the information about the 'stem' so far should be maintained. A simple list of characters would suffice for this purpose. As the traversal is proceeding, this list acts as an explicit stack, but each time a prefix or a complete word is to be printed out, this list should be treated as a queue.

6.15. The ADT Lexical Table can be implemented using a hash table HT [0.

Max]. As a hash function, the multiplicative hashing function using the golden ratio can be used to remove most of the biases in the input data.

The implementation of the TRIPLE_SET operations is mainly concerned with manipulating multi-lists with pointers.

6.16.

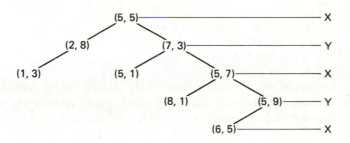

6.19. Briefly:

TYPE student:	A composed structure
TYPE course:	A composed structure
ADT courses:	set of course
ADT students:	set of student
ADT student_course:	many-to-many relationship

The set students requires sequential access based on student names and random access based on student numbers. This means that a B^+-tree cannot be used. For efficiency, two data structures should be used. A hashing table of pointers to student records can be used for random access (The size of the table can be calculated as $1000 + 10$ per cent of $1000 = 1100$.) For sequential access (for complete listings of students), a B^+-tree can be used which contains information about student names and a pointer to the actual student records. Of course a linear list could be used for this, but since information regarding students attending courses is subject to continuous change, this would be inefficient.

The set courses can be implemented as a vector of course records. To process 'all courses of a student' and 'all students in a course' a multi-list can be used. However, since the set courses is bounded by 50, a simpler approach is preferred. For each course record, a linked list of pointers to student records can be maintained. For each student record, a bit-map can be used to represent the set of courses a student takes.

```
bit_map = array[1..50] of boolean;
student = record
            student_no : integer;
            first_n, middle_n, last_name : packed array[1..15] of char;
```

```
        date_of_birth : record
                        day : 1..31;
                        month : 1..12;
                        year : 1900..1999
                      end;
        term_address, perm_address: packed array[1..128] of char;
          courses : bit_map
      end;
course = record
          name : packed array[1..20] of char;
          other : otherType
        end;
```

The operations. for this complex ADT include the operations on the set courses, set students and the many-to-many relationship between student and course objects.

Chapter 7

7.1   (a)   Nodes are procedures or functions of Pascal and an arc (p, q) indicates that p calls q.

       (b)   Nodes are junctions and arcs are roads between junctions. There may be alternative roads between two junctions.

7.2.   Take any node and designate that as the root of the tree. Then for any adjacent nodes to that node, designate them as children of that node and apply this process recursively for every such child.

7.4.   For sparse matrices, use a multi-list structure.

7.9.   Both of these problems are instances of the MCST problem.

7.15.   (c)   Modify the while loop to:

```
while not empty(frontier) and not member(T, permanent) do
```

       (b)   As for (c) but no need to compute T_x values; and the frontier can be implemented as a queue.

       (a)   As for (b) but a simple unordered list would suffice for the set frontier.

7.17.   (Also see chapter 8)

$$T_x = \begin{cases} 0 & \text{if } x = S \\[2ex] \underset{(y,\,x)\,\in\,E}{\text{minimum}} \, (T_y + \text{weight}(y,\,x)) & \text{otherwise} \end{cases}$$

In the worst case, to compute T_t, T_y must be computed for $n-1$ other nodes; for each of these T_y values, $n-2$ T_z must be computed, and so on. Therefore, in the worst case a time of $O(n!)$ is required.

7.18. This is primarily an application of the graph-searching algorithm (to determine if a path exists from a source to a destination).

7.20. A complete program to read a graph and to find the shortest weighted paths is given below. For details of the problem see section 7.4 on Dijkstra's algorithm (note there is no exercise **7.20**).

```
program shortpath(input, output);
  (* This program reads a directed graph and a source node s.
     Then, using Dijkstra's algorithm, it finds and reports
     the shortest weighted path lengths from s to all other
     nodes in the graph. *)
  const
    n = 15;  (* maximum number of nodes in the graph *)
  type
    node = 0..n;
    arc = record from_node, to_node : node;  weight : integer  end;
    vector = array[1..n] of node;
    zet = record  (* data structures to implement ADT set *)
             H, X : vector;          (*  - see section 7.4 *)
             size : 0..n
           end;
  var
    pred : array[1..n] of integer;
    T : array[1..n] of integer;
    graph : array[1..n,1..n] of integer;
      (* -maxint   indicates absence of an arc;
         -1        indicates presence of an arc;
         0..maxint indicates presence of an arc and
                             the weight of the arc *)

    s : node; (* the source node *)
    a : arc;

(*   A D T :   B O U N D E D   G R A P H - WITH N NODES  *)

  procedure init_graph;
    var i,j : integer;
    begin
      for i := 1 to n do
        for j := 1 to n do
          graph[i,j] := -maxint
    end;
procedure add_arc(a : arc);
  begin
    graph[a.from_node, a.to_node] := a.weight
  end;

procedure xset_node_pred(var n : node; e : node);
```

```
    (* e is the predecessor of n *)
    begin
      pred[n] := e
    end;

function get_pred(n : node) : node;
  begin
    get_pred := pred[n]
  end;

procedure set_node_label(var n : node; L : integer);
  (* L is T_n *)
  begin
    T[n] := L
  end;

function label1(n : node) : integer;
  begin
    label1 := T[n]
  end;

procedure xset_arc_nodes(var a : arc; n1,n2 :node);
  begin
    a.from_node := n1;
    a.to_node := n2;
    a.weight := -1
  end;

procedure set_arc_weight(var a : arc; weight : integer);
  begin
    a.weight := weight
  end;

procedure xprocess_nodes(procedure P(var n : node));
  var i, temp : node;
  begin
    for i := 1 to n do
      begin
        temp := i;  (* This is because the loop    *)
        P(temp)     (* variable cannot be changed. *)
      end
  end;

function edge(n1, n2 : node) : boolean;
  begin
    edge := graph[n1,n2] >= -1
  end;

function weight(n1, n2 : node) : integer;
  begin
    weight := graph[n1,n2]
  end;
  procedure process_adjacent_nodes(w : node;
                                      procedure P(var x : node;
                                                  var a : arc));

    var i : node; a : arc;
    begin
```

```
      for i := 1 to n do
        if graph[w, i] >= 0 then
          begin
            a.from_node := w; a.to_node := i;
            a.weight := graph[w,i]; P(i, a)
          end
    end;

  procedure adjacent_nodes(a : arc; var n1, n2 : node);
    begin
      n1 := a.from_node;   n2 := a.to_node
    end;

(* T H E    E N D    O F    A D T    G R A P H *)
(*   A D T    Z E T    *)

  function priority(x : node) : integer;
    begin
      priority := T[x]
    end; (* of priority *)

  procedure initialise(var z : zet);
    var i : integer;
    begin
      with z do
        for i := 1 to n do begin H[i] := 0; X[i] := 0 end;
      z.size := 0
    end;

  function empty(z : zet) : boolean;
    begin
      empty := (z.size = 0)
    end;

  procedure insert(x : node; var z : zet);
    var i, parent : integer; temp : integer;
    begin
      if (z.X[x] = 0) and (z.size < n) then  (* x is not in z and  *)
        begin                                (* set z not is full. *)
          i := z.size + 1;
          z.H[i] := x; z.X[x] := i; z.size := i;
          parent := i div 2;
          while (i > 1) and    (* the root has not yet reached *)
              (priority(z.H[i]) < priority(z.H[parent])) do
            begin
              (* Exchange elements i and parent *)
              temp := z.H[i];
              z.H[i] := z.H[parent]; z.X[z.H[i]] := i;
              z.H[parent] := temp; z.X[temp] := parent;
              i := parent;
              parent := i div 2
            end
        end
    end; (* of insert *)

  procedure delete_min(var x : node; var z : zet);
    var temp : integer; i,j : integer;
```

```
            more : boolean;
      begin
        if z.size = 0 then x := 0        (* Q is empty *)
        else
           begin
              x := z.H[1]; (* the element with smallest priority *)
              z.X[x] := 0;
              if z.size = 1 then z.size := 0
              else
                 begin
                    z.H[1] := z.H[z.size]; (* move the last element *)
                    z.size := z.size - 1;        (* to the root *)
                    z.X[z.H[1]] := 1;
                    (* Now push the first element into the heap *)
                    i := 1;
                    more := true;
                    while (i <= (z.size div 2)) and more do
                       begin
                          if (priority(z.H[2*i])
                              < priority(z.H[2*i+1])) or
                                (2*i = z.size) then
                                   j := 2 * i
                          else j := 2*i + 1;
                          (* j is the smaller child of i *)
                          if priority(z.H[i]) >
                                      priority(z.H[j]) then
                             begin
                                (* exchange these two elements *)
                                temp := z.H[i];
                                z.H[i] := z.H[j]; z.X[z.H[i]] := i;
                                z.H[j] := temp; z.X[temp] := j;
                                i := j
                             end
                          else              (* stop the process *)
                             more := false
                       end    (* of while *)
                 end
           end
      end;   (* of delete *)


function member(x : integer; z : zet) : boolean;
   begin
      member := (z.X[x] <> 0)
   end;


procedure modify_weight(x : integer; var z : zet);
   (* This procedure assumes that T_x has decreased as
      is the case in the Dijkstra's algorithm. Therefore,
      a variant of the insert algorithm is used below. *)
   var i, parent : integer; temp : integer;
   begin
      i := z.X[x];
      if i <> 0 then (* x is in the set z *)
         begin
            parent := i div 2;
            while (i > 1) and      (* the root has not yet reached *)
                  (priority(z.H[i]) < priority(z.H[parent])) do
```

```
               begin
                 (* Exchange elements i and parent *)
                 temp := z.H[i];
                 z.H[i] := z.H[parent]; z.X[z.H[i]] := i;
                 z.H[parent] := temp; z.X[temp] := parent;
                 i := parent;
                 parent := i div 2
               end
         end
   end; (* of modify_weight *)

(* T H E    E N D    O F    A D T    Z E T *)

 procedure dijkstra;
   var frontier : zet;
       w : node;
 procedure set_T(var x : node);
   begin
     set_node_label(x, maxint)
   end;

 procedure set_P(var x : node);
   begin
     xset_node_pred(x, 0);
   end;

 procedure expand(var x : node; var a : arc);
   var L : integer;
   begin
     L := (labell(a.from_node) + weight(a.from_node, a.to_node));
     if labell(x) > L then
       begin
         set_node_label(x, L);
         xset_node_pred(x, a.from_node);
         if not member(x, frontier) then
           insert(x, frontier)
         else modify_weight(x, frontier)
       end
   end;

 begin
   xprocess_nodes(set_T);
   xprocess_nodes(set_P);
   set_node_label(s, 0);
   initialise(frontier);
   insert(s, frontier);
   while not empty(frontier) do
     begin
       delete_min(w, frontier);
       process_adjacent_nodes(w, expand)
     end
 end; (* of dijkstra *)

 procedure read_graph;
   var f, t, w : integer;
       a : arc;
   begin
```

```
        writeln('There are ',n: 4,' nodes in the graph.');
        init_graph;
        writeln('Type the arcs and their weights ');
        writeln('one at a line in the form :-');
        writeln('FROM_NODE   TO_NODE   WEIGHT.');
        writeln('Terminate the input by typing a zero.');
        read(f);
        while f <> 0 do
          begin
            readln(t, w);
            if (w >= 0) and (f <= n) and (f >= 1)
                        and (t <= n) and (t >= 1) then
              begin
                xset_arc_nodes(a, f, t);
                set_arc_weight(a, w);
                add_arc(a)
              end
            else
              begin
                writeln('Nodes do not exist and/or weight is negative.');
                writeln('Try again.')
              end;
            read(f)
          end;
        readln
      end; (* of read_graph *)

procedure print_paths;
  procedure path(var n : node);
    procedure one_path(var n : node);
      var pred : integer;
          temp : node;
      begin
        pred := get_pred(n);
        if pred <> 0 then
            begin
              temp := pred;
              one_path(temp);
              write(' to ',n:4)
            end
          else
              write(n:4)
      end; (* of one_path *)

    begin
      if (n <> s) and (labell(n) <> maxint) then
          begin
            writeln('Path to node ',n:3,' with weighted path length ',
                                    labell(n):4,' : ');
            write('        ');
            one_path(n);
            writeln; writeln
          end
    end; (* of path *)

  begin
    xprocess_nodes(path)
```

```
     end; (* of print_paths *)
begin (* main program *)
  read_graph;
  (* read a source node *)
  writeln('Type the source node : 1 to ',n:3,'.');
  readln(s);
  dijkstra;
  print_paths
end.
```

The following is the output of the above program when applied to the graph G4 of diagram 7.10.

```
There are    15 nodes in the graph.
Type the arcs and their weights
one at a line in the form :-
FROM_NODE   TO_NODE   WEIGHT.
Terminate the input by typing a zero.
1 2 2
1 5 10
2 3 3
2 5 7
3 4 2
4 5 1
5 3 2
0
Type the source node : 1 to   15.
1
Path to node    2 with weighted path length    2 :
         1 to    2

Path to node    3 with weighted path length    5 :
         1 to    2 to    3

Path to node    4 with weighted path length    7 :
         1 to    2 to    3 to    4

Path to node    5 with weighted path length    8 :
         1 to    2 to    3 to    4 to    5
```

Chapter 8

8.2.

```
program permute(input, output);
  const x = 15;
  var s : array[0..x] of integer;
      i, N : integer;

  procedure print;
    (* to print one permutation *)
    var i : integer;
    begin
      for i := 1 to N do write(s[i],' ');
      writeln
    end;
```

```
procedure perm(i,c : integer);
  (* Find permutation of c+1..N by fixing c in the ith position *)
  var k : integer;
  begin
    s[i] := c;
    if c = N then print
    else
      for k := 1 to N do
        if s[k] = 0 then
          perm(k, c+1);
    s[i] := 0
  end; (* of perm *)

begin
  writeln('Type value of N <= ',x,'.');
  readln(N);
  for i := 0 to N do s[i] := 0;
  perm(0, 0)
end.
```

8.3.

```
procedure generate(n : integer);
  begin
    S[n] := true;
    if n = 1 then process_combination(S)
    else generate(n-1);
    S[n] := false;
    if n = 1 then process_combination(S)
    else generate(n-1)
  end;
......
generate(n);
```

8.4. Assuming that the variable values are stored in the array S, then:

```
function evaluate (t : ptr) : boolean;
  begin
    case t↑.tag of
      variable : evaluate := S[t↑.index];
      Lnot :     evaluate := not(evaluate(t↑.branch));
      Lbinop :
        case t↑.binop of
          Land:   evaluate := evaluate(t↑.left) and evaluate(t↑.right);
          Lor:    evaluate := evaluate(t↑.left) or evaluate(t↑.right);
          Limp: evaluate := not(evaluate(t↑.left)) or evaluate(t↑.right);
          Lequiv: evaluate := (evaluate(t↑.left) = evaluate(t↑.right))
        end;
    end;
  end;
```

8.5.

```
function satisfiable(t : ptr) : boolean;
   label 1;
   procedure generate(n : integer);
      begin
        S[n] := true;
        if n = 1 then begin if evaluate(t) then goto 1 end
        else generate(n-1);
        S[n] := false;
        if n = 1 then begin if evaluate(t) then goto 1 end
        else generate(n-1)
      end; (* of generate *)

   begin
     satisfiabile := true;
     generate(n);
     satisfiable := false;
1 : end; (* of satisfiable *)
```

8.7.   *Hint:* use the following configuration:

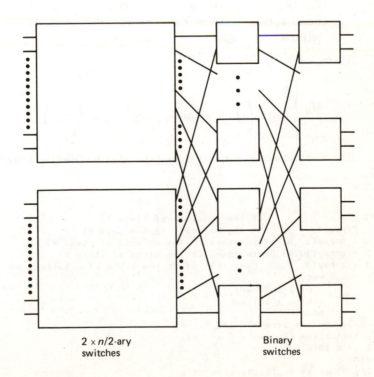

2 × *n*/2-ary   Binary
switches   switches

8.9. This is similar to finding the shortest path between two nodes in Dijkstra's algorithm in chapter 7.

8.14. M_{ij} = Minimum weighted path length of a tree constructed from $K_i K_{i+1} \ldots K_j$.

Divide and conquer:

The cost of this tree is

$$\underbrace{\sum (w(L+1))}_{\substack{\text{nodes in the} \\ \text{left tree}}} + \underbrace{\sum (w(L+1))}_{\substack{\text{nodes in the} \\ \text{right tree}}} =$$

$$\underbrace{\sum (wL)}_{\substack{\text{nodes in the} \\ \text{left tree}}} + \underbrace{\sum (wL)}_{\substack{\text{nodes in the} \\ \text{right tree}}} + \underbrace{\sum (w)}_{\substack{\text{nodes in the} \\ \text{left tree}}} + \underbrace{\sum (w)}_{\substack{\text{nodes in the} \\ \text{right tree}}}$$

Therefore

$$M_{ij} \begin{cases} 0 & i = j \\ \underset{i<x<j}{\text{MIN}} \left(M_{i,x-1} + M_{x+1,j} + \sum_{i}^{x-1} W_y + \sum_{x+1}^{j} W_y \right) & j > i \end{cases}$$

The time and space complexities will therefore be $O(n^3)$ and $O(n^2)$.

8.19.

```
const N = ;                 (* The total no of items *)
      Capacity = ;          (* the capacity of the sack *)
var W : array[1..N] of integer; (* The weights of items *)
    S : array[1..N] of integer; (* The sizes of items *)
    T : array[0..N, 0..Capacity] of (* The table of solutions *)
              record
                 wt : integer;
                 size : integer
              end;
(* ..Read the two arrays W and S.. *)
(* To initialise the table T :- *)
for i := 0 to N do
  begin
    T[i,0].wt := 0; T[i,0].size := 0
```

```
      end;
for i := 0 to Capacity do
  begin
    T[0,i].wt := 0; T[0,i].size := 0
  end;

(* Compute the array *)
for i := 1 to N do
  for j := 1 to Capacity do
    begin
      if (T[i-1, j].size + S[i] <= j) then
        begin
          T[i,j].wt := T[i-1,j].wt + W[i];
          T[i,j].size := T[i-1,j].size + S[i]
        end
      else if j - S[i] < 0 then
        T[i,j] := T[i-1, j]
      else if T[i-1, j-S[i]].wt + W[i] > T[i-1, j].wt then
        begin
          T[i,j].wt := T[i-1,j - S[i]].wt + W[i];
          T[i,j].size := T[i-1,j-S[i]].size + S[i]
        end
      else
        T[i,j] := T[i-1, j]
    end;
```

8.20. Dynamic programming produces the Pascal triangle:

```
1
1  1
1  2  1
1  3  3  1
1  4  6  4  1
. . . . . . . . . . . .
```

Chapter 9

9.2.

```
var data : array[1..N] of integer;
    aux  : array[1..N] of integer;  (* to merge the sublists *)
. . . . . . . . . .
procedure mergesort(L,U : integer); (* sort data[L..U] *)
  var i, j, k, m : integer;
  begin
    if U > L then
      begin
        m := (L + U) div 2;
        mergesort(L,m);
        mergesort(m+1, U);
```

```
(* Merge the two sublists.
   This corresponds to the procedure JOIN *)
i := L; j := m + 1; k := L;
while (i <= m) and (j <= U) do
   begin
      if data[i] < data[j] then
         begin
            aux[k] := data[i]; i := i + 1
         end
      else
         begin
            aux[k] := data[j]; j := j + 1
         end
      k := k + 1
   end (* of while *)
if i < m + 1 then
   for j := i to m do
      begin
         aux[k] := data[j]; k := k + 1
      end
else
   if j < U + 1 then
      for i := j to U do
         begin
            aux[k] := data[i]; k := k + 1
         end
   (* Now copy aux into data *)
   for i := L to U do data[i] := aux[i]
   end
end; (* of mergesort *)
```

9.4.   (i) $d = \log 16 + 1 = 5$, therefore there would be 5 sub-lists numbered 1 to 5:

List   25 2 36 142 101 84 32 49   8 321   25 458 981 122   48   95
Sub-lists 1 2 3   4   5 1 2 3   4   5   1   2   3   4   5   1

Each sub-list should be sorted using an insertion sort:

25 2 36   8 48 25 32 49 122 101   84 485 981 142 321   95

(ii) Now, choose $d = 5$ div $2 = 2$, therefore we get two sub-lists:

List   25 2 36   8 48 25 32 49 122 101   84 485 981 142 321   95
Sub-lists 1 2 1   2   1 2 1 2 1   2   1   2   1   2   1   2

These two sub-lists are sorted:

25 2 32   8 36 25 48 49   84   95 122 101 321 142 981 485

(iii) Finally, $d = 2$ div $2 = 1$; that is, there will be only one sub-list to sort:

25 2 32   8 36 25 48 49   84   95 122 101 321 142 981 485

sorted to

2 8 25   25   32 36 48 49   84   95 101 122 142 321 485 981

9.11.   O(n) in the worst case and O($\log n$) in the best and average cases.

9.13.
```
procedure push (L, U: integer);
  var i, j, save : integer; finished : boolean;
  begin
    i := L;
    j := 2 * L;
    save := A[L];
    finished := false;

    while (j <= U) and not finished do
      begin
        if j < U then
          if A[j] > A[j+1] then
            j := j + 1;
        if save <= A[j] then
          finished := true
        else
          begin
            A[i] := A[j];
            i := j;
            j := 2*i
          end
      end;
    A[i] := save
  end;
```

9.14. Let $S(i,j) = \sum\limits_{x=i}^{j} \dfrac{x}{2^x}$

$$S(1,n) = 1/2 + S(2,n)$$
$$= 1/2 - (n+1)/2^{n+1} + S(2,n+1)$$

$$= 1/2 - (n+1)/2^{n+1} + 1/2\, S(1,n) + 1/2 \sum_{i=1}^{n} 1/2^i$$

$$1/2\, S(1,n) = 1/2 - (n+1)/2^{n+1} + 1/2 \sum_{i=1}^{n} 1/2^i$$

Therefore $S(1,n) = 2 - (n+2)/2^n$. *Q.E.D.*

9.15. To get the first k elements of a list takes a time of $O(n)$ for $k \leqslant n/\log n$.

Chapter 10

10.2. Note that each cycle should be reported only once and that the cycle (a, b, c, d, a) is the same as the cycle (d, a, b, c, d). The solution for this, in fact involves examining all possible cycles of the graph and in the worst case it needs a time of $O(n!)$ to find and report the output. Also, see solutions to exercises **10.9** and **10.10** below.

10.6. The data structures for the matching problem were discussed in chapter 10. The following program reads a bipartite graph and finds a matching for it.

```
program matching(input, output);
  const
    N = 30;  (*  Maximum no of nodes in the graph  *)
  type
    node = 1..N;
    eptr = ↑edge;
    node_cell = record
                  matched : boolean;
                  first : eptr
                end;
    edge = record
             matched : boolean;
             i, j : integer;  (*  head and tail  *)
             next_i, next_j : eptr;
           (*  to next edge (i,_) and (j,_) respectively  *)
           end;
    cellptr = ↑cell;
    cell = record
             edge : eptr;
             next : cellptr
           end;
    list = cellptr;
  var
    nodes : array [node] of node_cell;
    edges : eptr;
    G, M : LIST;            (*  LIST of ↑edge  *)
    V1p, V2p : 1..N;
    p : eptr;
    i, j, no1, no2 : integer;
    found : boolean;
    L : list;
procedure initialise_list (var L : list);
  begin
    new(L);
    L↑.next := nil
  end;  (*  of initialise_list  *)

procedure insert_list (p : eptr; var L : list);
  var
    t : cellptr;
  begin
    new(t);
    t↑.edge := p;
    t↑.next := L↑.next;
    L↑.next := t
  end;  (*  of insert_list  *)

procedure XOR (var M : LIST; G : LIST);
  var
    ml, gl : list;
    p : eptr;
    found : boolean;
  begin
    gl := G;
    while gl↑.next <> nil do
```

```
    begin
      found := false;
      ml := M;
      while (ml↑.next <> nil) and not found do
        if ml↑.next↑.edge = gl↑.next↑.edge then
          found := true
        else
          ml := ml↑.next;

      if found then
        begin
          ml↑.next↑.edge↑.matched := false;
          (* This node can be disposed of *)
          ml↑.next := ml↑.next↑.next;
          if ml↑.next <> nil then
            ml := ml↑.next;
        end
      else
        begin
          p := gl↑.next↑.edge;
          p↑.matched := true;
          nodes[p↑.i].matched := true;
          nodes[p↑.j].matched := true;
          insert_list(p, M);
        end;
      gl := gl↑.next;
    end;
  end;  (*  of XOR  *)

procedure  dfs (i, L : integer; var found : boolean);
  var
    p : eptr;
  begin
    found := false;
    if not odd(L) and not nodes[i].matched then
      found := true
    else
      begin
        p := nodes[i].first;
        while not found and (p <> nil) do
          begin
            if p↑.i = i then
              j := p↑.j
            else
              j := p↑.i;
            if (p↑.matched and not odd(L))
                        or (not p↑.matched and odd(L)) then
              begin
                dfs(j, L + 1, found);
                if found then
                  insert_list(p, G)
              end;
            if p↑.i = i then
              p := p↑.next_i
            else
```

```
                     p := p↑.next_j
            end
        end
   end;  (* of DFS *)


begin
  (*  READ THE BIPARTITE GRAPH  *)
   writeln('Type the No of nodes in V1 and V2 respectively.');
   writeln('Note that V1 + V2 must not exceed', N : 3, '.');
  readln(no1, no2);
  (*  no1 in the number of nodes in V1 and  *)
  (*  no2 in the number of nodes in V2  *)
  v1p := no1;                  (*  Nodes 1..v1p are in V1  *)
  v2p := N - no2 + 1;       (*  Nodes v2p..N are in V2  *)


for i := 1 to N do
   nodes[i].first := nil;


writeln('Now type the arcs as pairs (x,y) such that x is in');
writeln('V1 and y is in V2.  The pairs should be typed one on a');
writeln('line and terminated by a zero on the last line.');
read(i);
while i <> 0 do (* Read the arcs (i,j) such that i is in V1 and *)
   begin               (*         and j is in V2                    *)
      readln(j);
      (* Now add the edge (i,N-j+1) to the graph *)
      new(p);
      p↑.i := i;
      p↑.j := N - j + 1;
      p↑.matched := false;
       if nodes[i].first = nil then
         begin
            nodes[i].first := p;
            p↑.next_i := nil
         end
       else
         begin
            p↑.next_i := nodes[i].first;
            nodes[i].first := p
         end;
       if nodes[N - j + 1].first = nil then
         begin
            nodes[N - j + 1].first := p;
            p↑.next_j := nil
         end
       else
         begin
            p↑.next_j := nodes[N - j + 1].first;
            nodes[N - j + 1].first := p
         end;
      read(i);
   end;
   readln;
```

```
(*  FIND A MAXIMAL MATCHING  *)
INITIALISE_LIST(M);  (*  M will be the maximal matching  *)
(*  For all e in E do e.matched := false  *)
for i := 1 to N do   (*  For all i in V do i.matched := false  *)
  nodes[i].matched := false;

for i := 1 to vlp do
  if not nodes[i].matched then
    begin
      INITIALISE_LIST(G);
      dfs(i, 1, found);
      if found then
        XOR(M, G);
    end;
  L := M;
  writeln('The maximal matching consists of the arcs :-');
  while L↑.next <> nil do
    begin
      writeln(L↑.next↑.edge↑.i : 5, N - L↑.next↑.edge↑.j + 1 : 5);
      L := L↑.next
    end
end.
```

10.8. If we follow the example of the chapter 3 where we read an expression in Reverse Polish and constructed its binary tree representation, we can proceed as:

  (i) Transform the expression into Reverse Polish.

  (ii) Use a stack of ptr where ptr is a pointer to a tree to construct the tree, with the difference that every time a tree pointer is pushed into the stack, the trees in the stack must be searched to find if this tree is already present; if it is, then a pointer to this tree will be pushed into the stack instead. Note that we actually need a list for this purpose since elements are added and deleted in the list in an LIFO fashion, but they are searched in a sequential manner. For the above example, the list would grow and shrink as

     ab*23ab*/−+

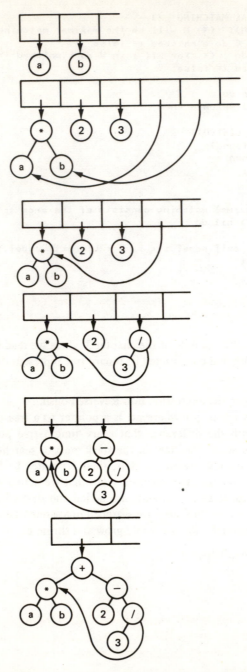

10.9. One might think that the cycle-detection algorithm can be used here, since the length of the current cycle being found can be maintained, and if it is equal to *n* then there is a Hamilton cycle. But, this does not work, since our algorithm only gives us sub-cycles of a Hamilton cycle (if there is one) and not the main cycle. For example, the dfs of the following graph gives us the cycle (c, d, b, c), but does not indicate that the cycle (c, d, a, e, b, c) also exists.

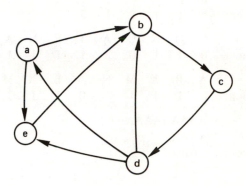

For this problem, all possible paths must therefore be investigated. This means that a node must be searched, even though it has been visited before:

```
count := 0;
for all v in V do csp[v] := false;
(* This is to stop going round a cycle indefinitely *)
dfs(x);        (* x is an arbitrary node of the graph *)

where :-

dfs(v) :
  begin
    csp[v] := true;
    count := count + 1;
    for all (v,w) in E do
      if (count = n - 1) and (w = x) then A Hamilton cycle found
      else if not csp(w) then dfs(w);
    csp[v] := false;
    count := count - 1
  end;
```

This algorithm needs $O(n!)$ in the worst case. This is because, in the worst case, node x has $n-1$ adjacent nodes (say y); each node y has $n-2$ adjacent nodes, and so on. This gives $n \times n-1 \times n-2 \times \ldots \times 1$ different cases.

10.10. This problem is again similar to the Hamilton cycle problem of exercise **10.9** and similar arguments apply. The time complexity would be $O(n!)$ in the worst case.

Chapter 11

11.4.

pattern:	A	B	C	A	C	A	C	D	F	A			
fail function:	0	1	1	0	2	0	2	1	1	0			

pattern:	C	O	N	C	O	C	T						
fail function:	0	1	1	0	1	3	2						

pattern:	C	O	N	C	A	T	E	N	A	T	E		
fail function:	0	1	1	0	2	1	1	1	1	1	1		

pattern:	A	B	X	M	N	A	B	X	M	N	A	B	B
fail function:	0	1	1	1	1	0	1	1	1	1	0	1	8

11.5.

text:	AAAAAAAAAAA
pattern:	AAAX
fail function:	0003

Chapter 12

12.1. Naturally $F(n) \geqslant F(n-1)$.

Min cost for TSP $\leqslant 1/2 \sum_{i=1}^{n}$ the two arcs incident on node i with max weights

$\leqslant m$

Now, using a binary search operation:

if there exists a tour with cost $\leqslant m/2$ then
    repeat search for cost $\leqslant m/4$
else repeat search for cost $\leqslant 3m/4$

Therefore, the binary search gives the minimum cost of a tour (C say) of the graph G in time proportional to $(F(n) \times \log m)$ (max).

To get the actual minimal cost tour, we can proceed as follows:

- Take node v; and construct graph G' by removing node v and all the arcs incident on it in G

- Find C', the minimum cost of G', $C' \leqslant F(n-1) \times \log m'$
$\leqslant F(n) \times \log m$

- Choose two arcs a1 and a2 such that:

$\quad$ weight(a1) + weight(a2) $-$ weight(a3) + C' = C where
$\quad$ a1 = (v, x) $\quad$ a2 = (v, y) $\quad$ and $\quad$ a3 = (x, y) in G'

This takes $O(n)$ since G' has $n-1$ nodes.

- Add a1 and a2 to the solution.

The time for this recursive operation is bounded by $n(n + F \times \log m)$. Therefore, the total time for finding the optimum solution is

$$n(n + F \times \log m) + F \log m = n^2 + [(n+1) \log m] \times F$$
$$= P(F)$$

where P is the polynomial $P(a) = n^2 + [(n+1) \log m] \times a$.

12.2.   The principle of optimality holds for the TSP. If optimal tours for graphs with n' nodes and $n-n'$ nodes (disjoint) are known, we can construct an optimal tour for the entire graph by taking all pairs of such sub-graphs and trying to paste them together. This pasting operation can be done in $O(n)$. However the total number of sub-problems to solve and also the total number of distinct sub-problems will remain exponential, and therefore the resulting algorithm will not be polynomial (although it will be better than the exhaustive search algorithm).

n problems of size 1 node
C_2^n problems of size 2 nodes
C_3^n problems of size 3 nodes
. . . .
C_n^n problem of size n nodes

The total number of the sub-problems T is $\sum\limits_{i=1}^{n} C_i^n$

$$\sum_{i=0}^{n} a^n b^{n-1} C_i^n = (a+b)^n$$

$$T = (1+1)^n - 1 = 2^n - 1$$

12.21. Construct a graph $G=(V, E)$ where V represents the members of S and E contains arcs (v, w) if $(v, w) \in R$. In the maximal complete sub-graph of G, each node must be mapped to a unique value (in the range 1 to k). Other nodes not in this sub-graph can be mapped to values 1 to k since, if there is a contention, there needs to be a complete sub-graph in G with m nodes where $m > k$.

Index

374